AF538365

The Process Evaluation Handbook

Donald J. Wheeler

SPC Press
Knoxville, Tennessee

SPC Press
5908 Toole Drive, Suite C
Knoxville, Tennessee 37919
(865) 584–5005
Fax (865) 588–9440

ISBN 0–945320–55–8

xii + 250 pages
58 figures
45 tables
4 Worksheets

1 2 3 4 5 6 7 8 9 0

Contents

Tables 1 to 20

The Effective Cost of Production for Two-Sided Specifications **97 - 158**

About the Author

Donald J. Wheeler is a continual improvement specialist who had the great good fortune to work with Dr. W. Edwards Deming and David S. Chambers. Dr. Wheeler graduated from the University of Texas, Austin, with a Bachelor's degree in Physics and Mathematics, and holds M.S. and Ph.D. Degrees in Statistics from Southern Methodist University. From 1970 to 1982 he taught in the Statistics Department at the University of Tennessee, Knoxville, where he was Associate Professor. Between 1981 and 1993 he periodically assisted Dr. Deming with his four-day seminars. He is author or co-author of 19 books and over 130 articles, and is a Fellow of the American Statistical Association.

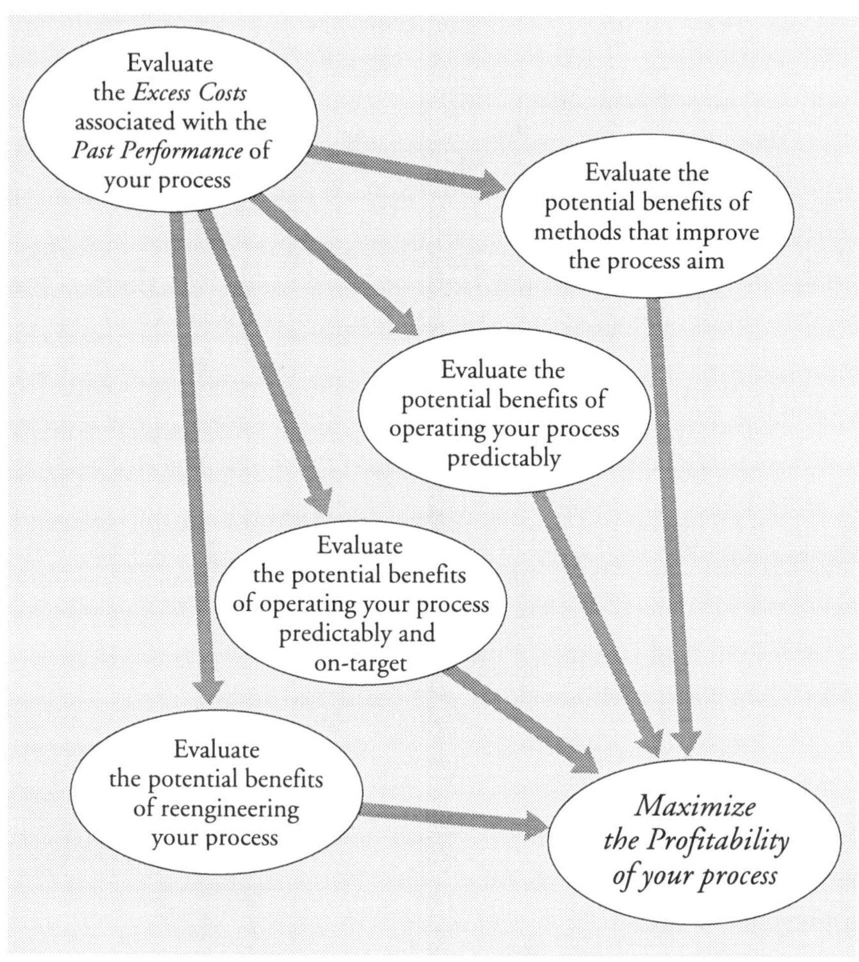
Evaluate
the *Excess Costs*
associated with the
Past Performance of
your process
Evaluate the
potential benefits of
methods that improve
the process aim
Evaluate the
potential benefits of
operating your process
predictably
Evaluate
the potential benefits
of operating your process
predictably and
on-target
Evaluate
the potential benefits
of reengineering
your process
Maximize
the Profitability
of your process

Preface

In any journey it is important to know two things—where you want to go and where you are starting from. Until these two things are known, it will be difficult to plot a course to get from one to the other. This handbook is intended as a guide for planning your course. It is intended as an aid for those who have the responsibility for maintaining and improving production processes.

This is not a book about capability indexes—it is a book about how to convert capability indexes into dollars. This is not a book about process control schemes—it is a book about how much you might save with a process control scheme. This is not a book about process behavior charts—it is a book about the dollars you can save by operating your process predictably. As the name implies, this book is about process evaluation—where you have been, where you can go, and what you can accomplish by going there. Moreover, this assessment is done using the language of management—the dollar savings that can be realized from a given course of action.

But which course of action? Should you improve the process aim? Or should you seek to operate the process predictably? Or should you try to do both of these by operating predictably and on-target? What about a process upgrade or even reengineering the whole process? The process evaluation techniques given here will help you to estimate the payback for each of these different approaches to improvement. They will define the potential savings for each of these courses of action before you actually undertake a given course of action. By choosing those projects with the larger payback potential you will have the maximum impact with the least effort.

So, regardless of the approach to improvement you may be using, this handbook will help you to choose the right projects, and once embarked, it will help you to choose the right course of action to achieve the objective of maximizing the profitability of your process.

Introduction

Whenever we evaluate a process we inevitably have to discuss specifications. Unfortunately, specifications are a disparate group of things sharing a common label. As a result, there is some confusion surrounding the nature and purpose of specifications.

I will use the label of *product specifications* to refer to the most common type of specifications. Product specifications define those points where we take action on the product. Actions that keep the nonconforming product from being used for its originally intended purpose are equivalent to scrapping the product. Actions that allow the nonconforming product to eventually be used for its intended purpose will be called rework. "Specifications" that do not correspond to points where you take action to separate the acceptable product from the unacceptable product are not product specifications.

Another category of specifications are *process specifications*. These are specifications where the action is not taken on the individual process outcomes, but rather is applied to the process itself. These specifications often take the form of some percentage conforming or nonconforming. And when this percentage takes a turn for the worse it is time to change the process.

A third category of specifications are *safety specifications*. These are specifications that allow some regulatory agency to take legal action against the producer. These specifications usually get a very wide berth.

In this book we shall be concerned with the first two categories of specifications. Product specifications will be the focus of Chapters One through Six, and process specifications will be the topic in Chapter Seven.

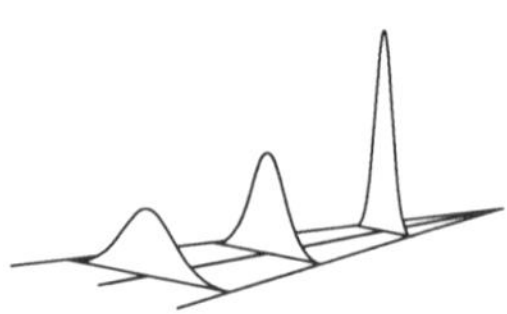

Chapter One

The Effective Cost of Production

1.1 The Traditional View of Excess Production Costs

The fraction of nonconforming product has traditionally been used to track production because nonconforming product always creates excess production costs—you lose the cost of production for every item that you have to scrap, and you incur an additional cost for every item that you rework. Ten percent scrap means that ten percent of your production costs were wasted. Fourteen percent rework means that you have incurred an extra expense for fourteen percent of your product. And of course there is the additional expense of inspection to determine which parts to scrap and which parts to rework.

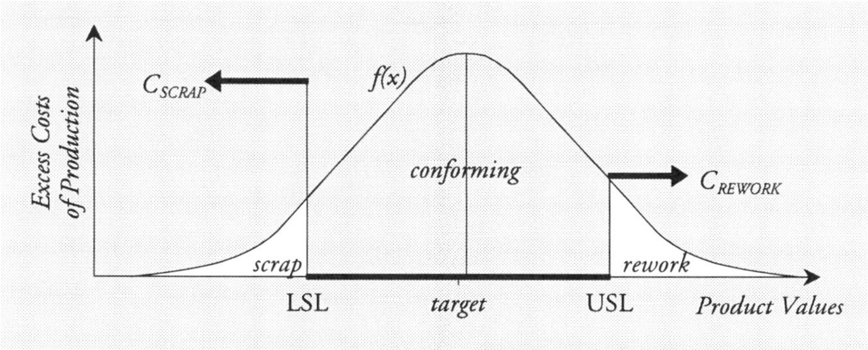

Figure 1.1: The Traditional View of Excess Costs of Production

These excess production costs are easy to identify. Graphically they could be characterized as in Figure 1.1. When the product value falls below, say, the lower specification limit the manufacturer will incur the costs of scrap. When the product value falls above the upper specification limit the manufacturer will incur the cost of rework. And we traditionally assume that the conforming items do not incur any excess costs. Under these conditions the excess cost function of Figure 1.1 could be expressed as:

$$\begin{aligned} \textit{excess costs} &= \textit{cost of scrap} \times \textit{proportion scrapped} \\ &+ \textit{cost of rework} \times \textit{proportion reworked} \end{aligned}$$

where the proportion scrapped is the area under the curve labeled $f(x)$ that falls below the lower specification limit, and the proportion reworked is the area under the $f(x)$ curve that falls above the upper specification limit. If, as shown in Figure 1.1, you had 10 percent scrap, and about 14 percent rework, and the cost of rework, C_R, was one-half the cost of scrap, C_S, then you would have an excess cost of:

$$\textit{excess cost} = 0.10\ C_S + 0.14\ C_R = (0.10 + 0.07)\ C_S = 0.17\ C_S$$

If we consider the cost of scrap to be essentially the same as the nominal cost of production, then our excess costs amount to 17 percent over the nominal cost of production. However, since scrap deflates our output, our actual cost per unit shipped would be found by adding the nominal cost of the units shipped to the excess costs for the units produced, and then dividing by the number of units shipped:

$$\begin{aligned} \textit{actual cost} &= \frac{\textit{nominal cost of units shipped} + \textit{excess costs of units produced}}{\textit{units shipped}} \\ &= \textit{nominal cost per unit} \times \frac{1.0 - \textit{proportion scrapped} + \textit{excess costs}}{1.0 - \textit{proportion scrapped}} \\ &= \textit{nominal cost per unit} \times \textit{Effective Cost of Production} \end{aligned}$$

And the Effective Cost of Production provides you with a yardstick on how efficient your production operation actually is. For the example above ten percent were scrapped and the excess cost per unit was 0.17, giving:

$$\textit{Effective Cost of Production} = \frac{1.0 - 0.10 + 0.17}{0.90} = \frac{1.07}{0.90} = 1.189$$

which we would interpret to mean that our actual cost per unit produced was 1.189 times the nominal cost per unit. Thus, in this case, 10 percent scrap and 14 percent rework turns into a 19 percent increase in the cost of producing each and every unit shipped. We will return to this calculation, but first we need to examine, and revise, the model used for the excess costs.

The traditional model for the excess costs of production is the one shown in Figure 1.2. According to this model, if there is no scrap, and if there is no rework, then there will be no excess costs of production. Unfortunately, life is not quite this simple.

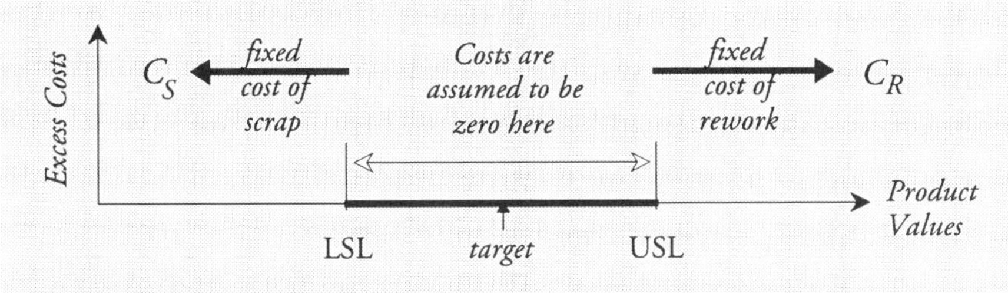

Figure 1.2: The Traditional Model for Excess Costs of Production

The traditional model for excess costs of production is a simplification—while it takes into account our overt actions, it ignores the reasons for those actions. The original idea behind mass production was to make every item alike. We quickly learned that we could not make every item exactly the same, so we had to settle for making them similar. And then we had to decide how similar was similar. Obviously a part that was right at the target value would be considered to be a good part. And parts that were very close to the target value would still be functional. But as a part would deviate from the target value by greater and greater amounts we could see that the functionality of that part would, of necessity, suffer. The greater the deviation, the greater the loss in functionality. And finally, at some point the part would become so nonfunctional that it would be unacceptable. At that point we would draw the line, call it the specification limit, and decide to either scrap or rework those parts that cross over the line. It is this loss in functionality which is inherent in deviations from the target value that is the primary motivation for specification limits.

But the excess cost function shown in Figure 1.2 does not reflect this reality. There the excess costs were assumed to be zero everywhere within the specifications. While this assumption may seem to cover the direct costs of production, it essentially denies the reality that specifications were designed to deal with in the first place—the reality that functionality suffers as parts deviate from the target value. While the indirect costs of decreased functionality may only show up subsequent to fabrication, they are still real costs, and eventually the fabricator will have to answer for these costs. Therefore, we need a more realistic model for the excess costs of production.

1.2 A More Realistic View of Excess Production Costs

The loss in functionality that occurs when an item deviates from the target value implies that there are excess costs associated with using conforming product. These are in addition to the excess costs of scrap and rework. Yet, in practice, we have commonly ignored the costs of using conforming product. We have forgotten that the original idea was to produce identical parts—each part just like the previous part, and all the parts at the target value. So we accept the fact that variation happens, and then proceed to take actions to deal with that variation after the fact.

What are these actions? Everyone who has ever been close to a production line knows about the headaches caused by conforming items that will not assemble. Part A is in spec, and Part B is also in spec, but they will not assemble, even with the help of the rubber hammer issued to every operator. (One time a foreman proudly showed me an operation that was essentially nothing more than blacksmithy—he was proud because he had cut the number of hammer-wielders in half!)

Another of the costs associated with using conforming product is the practice of presorting the components into categories to minimize the misfits at assembly, or sorting so that the units can be used in matched pairs. Other costs come from adjustments to the process settings to account for variation in the incoming product, on-line testing of assemblies, and the rework of assemblies that fail the on-line test. All of these actions, and their excess costs, are due to variation. These costs are real, and they do affect the bottom line even though we bury them in different places and in different ways. While we may have become so used to these excess costs that we consider them to be a routine part of production, they are actions that would not be necessary if every item was equal to the target value.

Think about how your process would run if all of your incoming components were completely uniform—each part just like the previous part, and all parts at the target. The discrepancy between your actual experience and this idealized process is due to variation, and actions to deal with variation after the fact will inevitably increase the cost of using the conforming product.

"But these costs are costs incurred downstream—I don't have to worry about them." While it is true that these costs of variation are not directly incurred at fabrication, they do affect the final product and eventually get noticed by the consumer. When this happens all of these excess costs will come back to haunt you in the form of reduced business. Therefore, any realistic assessment of the costs of production should incorporate these costs of *using* conforming product.

To do this we shall have to modify the cost function of Figure 1.2. If we critically regard the elements of the cost function in Figure 1.2 we will quickly decide that the cost of scrapping an item is a real cost. It is also one that is fairly easy to attach a number to. Thus, the horizontal line on the left is probably one of the more solid portions of the cost function.

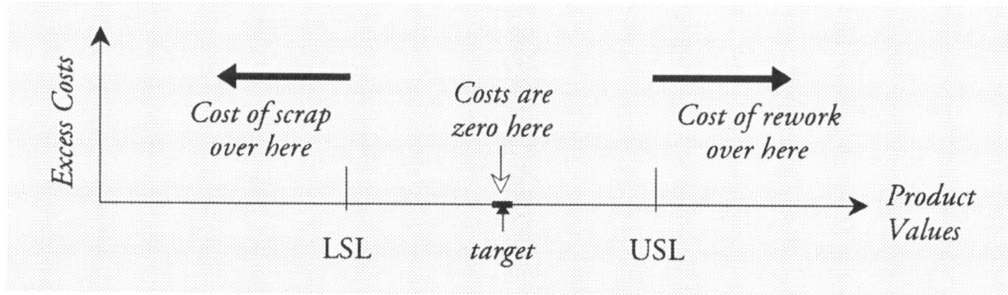

Figure 1.3: What We Really Know About the Costs of Production

Likewise, the cost of rework is probably one that can be worked out. In many plants this number is buried behind a departmental wall (as when a separate department is responsible for rework), but with a little effort a value can usually be found for this cost. Therefore, that portion of the cost function represented by the horizontal line on the right is fairly well known.

But where are the excess costs zero? The group of items that will logically have zero excess costs will be those items located right at the target value. (Excess costs may be defined as costs of actions incurred because items deviate from the target.) Thus, those parts of the cost function that are well-known are those shown in Figure 1.3.

A comparison of Figures 1.2 and 1.3 reveals that the problem with the traditional cost function is the *assumption* that the excess costs for conforming product are zero everywhere between the specification limits. It is reasonable to assume that the costs will be zero when the product value is very close to the target, as shown in Figure 1.3, but it is not reasonable to extend this zero cost to cover the whole of the region between the specification limits as was done in Figure 1.2.

So how can we turn the disjoint costs of Figure 1.3 into a realistic cost function? All of our experience with the physical world tells us that continuous functions provide more realistic models than do step functions. In addition, our experience tells us that there are costs associated with using *conforming* product. Thus, we need a reasonable way to connect the three "facts" shown in Figure 1.3. The simplest smooth curve that can be made to fit three points is a quadratic curve like the one shown in Figure 1.4.

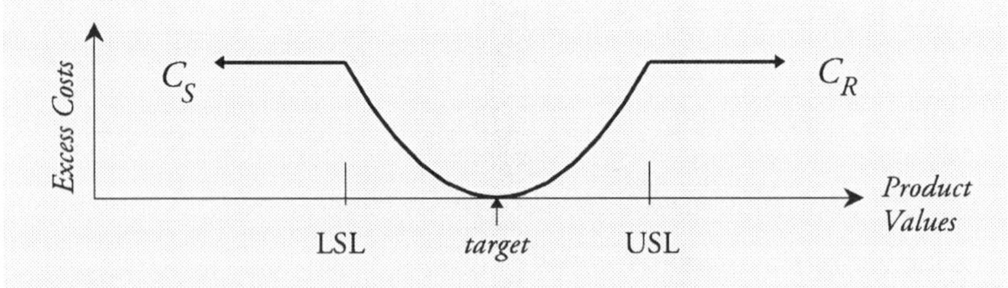

Figure 1.4: A More Realistic Model for the Excess Costs of Production

This is the origin of a more realistic model for the excess costs of production. It combines the costs due to scrap and the costs due to rework with the costs of variation about the target value. It is true to that underlying reality which causes us to take action to cut our losses at the specification limits—the fact that the functionality of the product decreases as we deviate from the target value—while it acknowledges the fact that small deviations from the target do not affect the functionality of the product. The inclusion of the cost of using conforming product into this model for excess costs may seem to broaden the definition of production, but it is important to note that scrap and rework are also manifestations of the customer requirements.

Now, how can we make use of this more realistic model? By combining it with a probability model of the process outcomes to obtain the excess costs for a given process.

1.3 The Mathematical Expression of Excess Costs

The excess cost function defines an excess cost for each possible product value. A probability model, *f(x)*, can be used to define the frequency with which each possible product value will occur. And so the average of the excess costs can be written as the integral of the product of the excess cost function and the probability model, *f(x)*.

$$\text{Excess Costs} = \int \text{excess cost function} \times \text{probability model}$$

Since our excess cost function comes in four parts the expression above separates out into four separate integrals as shown on the next page. There C_S will represent the cost of scrap and C_R will be the cost of rework.

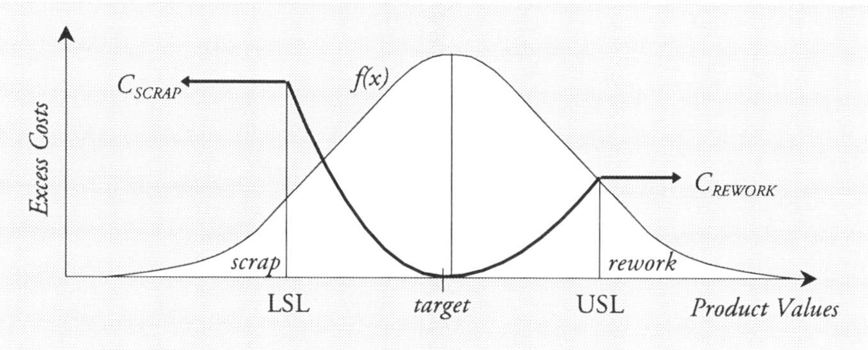

Figure 1.5: A More Realistic Model for the Excess Costs of Production Combined With a Normal Probability Model

$$\text{Excess Costs} = C_S \int_{-\infty}^{LSL} f(x)\,dx \qquad \text{excess costs due to scrap}$$

$$+\ C_S \int_{LSL}^{target} \frac{(x - target)^2}{(LSL - target)^2} f(x)\,dx \qquad \text{due to deviations below target}$$

$$+\ C_R \int_{target}^{USL} \frac{(x - target)^2}{(USL - target)^2} f(x)\,dx \qquad \text{due to deviations above target}$$

$$+\ C_R \int_{USL}^{\infty} f(x)\,dx \qquad \text{excess costs due to rework}$$

If we replace the four integrals with the following symbols:

$$\mathbf{ISP} = \int_{-\infty}^{LSL} f(x)\,dx \qquad \text{integral for scrap proportion}$$

$$\mathbf{IBT} = \int_{LSL}^{target} \frac{(x - target)^2}{(LSL - target)^2} f(x)\,dx \qquad \text{integral below target}$$

$$\mathbf{IAT} = \int_{target}^{USL} \frac{(x - target)^2}{(USL - target)^2} f(x)\,dx \qquad \text{integral above target}$$

$$\mathbf{IRP} = \int_{USL}^{\infty} f(x)\,dx \qquad \text{integral for rework proportion}$$

we can simplify the Excess Costs expression to become:

$$\begin{aligned} \textit{Excess Costs} &= C_S\,\mathbf{ISP} + C_S\,\mathbf{IBT} + C_R\,\mathbf{IAT} + C_R\,\mathbf{IRP} \\ &= C_S \left\{ \mathbf{ISP} + \mathbf{IBT} + \frac{C_R}{C_S}\left[\mathbf{IAT} + \mathbf{IRP} \right] \right\} \\ &= C_S \times \textit{Curly Brackets} \end{aligned}$$

In preparing the tables it was the Curly Brackets expression that was evaluated. The relationship between the process outcomes and the specifications is fully captured by the Curly Brackets term. The only additional value needed to completely specify the excess costs for a process is the cost of scrapping an item, C_S.

In order to fit this expression for excess costs into the earlier expression for the actual costs of production we need to factor out the nominal cost of production, *NCP*, giving:

$$\textit{Excess Costs} = NCP \left[\frac{C_S}{NCP} \times \textit{Curly Brackets} \right]$$

If we assume that the cost of scrap is essentially the same as the nominal cost of production, we end up with:

$$\textit{Excess Costs} = NCP \times \textit{Curly Brackets}$$

Now recall from page 2 that:

$$\begin{aligned} \textit{actual cost of production} &= NCP \times \textit{Effective Cost of Production} \\ &= NCP \times \frac{1.0 - \textit{proportion scrapped} + \textit{Curly Brackets}}{1.0 - \textit{proportion scrapped}} \end{aligned}$$

So that the Effective Cost of Production turns out to be:

$$\textit{Effective Cost of Production} = \frac{1.0 - \textit{proportion scrapped} + \textit{Curly Brackets}}{1.0 - \textit{proportion scrapped}}$$

Values of this expression are given in the tables. The different tables use different excess cost functions, or have the process average on different sides of the target value. Of special interest is the case where there is no scrap. In this case the expression above reduces down to:

$$\textit{Effective Cost of Production} = 1.00 + \textit{Curly Brackets}$$

You may use this fact to isolate the value of the Curly Brackets expression from Table 2 to use in custom computations.

In most of the tables the values are given to two decimal places. This was done because excessive precision is not necessary to evaluate the potential benefits of process improvements.

The tables were computed using a normal distribution for $f(x)$. While there are many other probability models that could be used to represent a predictable process, the normal distribution is a sufficiently general model to provide us with useful values for the Effective Cost of Production. Since the objective is to make comparisons between different Effective Costs of Production, approximate values will be sufficient to help you to make informed decisions about your process. While you could easily use some other model in the equations above, the Effective Costs of Production would not change appreciably, and the *differences* between the Effective Costs of Production for various situations, which are the numbers of greatest interest, would change even less.

Finally, while Figure 1.5 shows scrap on the left and rework on the right, the symmetry provided by the normal distribution makes the results in the tables completely general. Therefore, rather than left and right, the tables are presented in terms of "the scrap side of the target" and "the rework side of the target." In order to help you identify the situation covered by a given table, each table has its own graphic showing the excess cost function used and the relationship between the process average and the target.

1.4 The Tables of The Effective Cost of Production

The Effective Cost of Production incorporates the excess costs due to scrap, the excess costs due to rework, and the excess costs due to variation about the target but within the specifications. These excess costs will depend upon the scrap rate, the cost of scrap, the rework rate, the cost of rework, and the relationship between the product distribution and the specification limits. This approach to computing the Effective Cost of Production not only gives a more complete accounting, but it is also more realistic than the traditional calculation.

To save you the details of actually performing these computations this book contains tables of the Effective Costs of Production for various situations. Most of these tables use the common capability indexes, C_p and C_{pk} as descriptive values to allow you to find an Effective Cost of Production.

To use these tables you would begin by selecting an excess cost function that characterizes your process. For example, in Figure 1.4 the excess cost function shown is symmetric. This symmetric cost function is a reasonable default condition to use when you have two-sided specifications. However, in Figure 1.5 the excess cost function shown is not symmetric—it has a cost of rework that is one-half of the cost of scrap. When you know the relationship between these two costs then it makes sense to use this knowledge in selecting a excess cost function. (Since it does not make sense to rework an item when it is cheaper to scrap that item, I have assumed that the cost of rework will always be less than the cost of scrap.)

Once you have selected an excess cost function you can then use the capability indexes to obtain an Effective Cost of Production for your process. The values in the tables are expressed as multiples of the nominal cost of production, therefore a value of 1.28 would mean that your Effective Cost of Production was 28 percent greater than your nominal cost per unit. Of course, one of the major uses of values such as these is to allow you to compare the costs of production for different situations.

1.5 The Process Evaluation Procedures

Chapters Two through Six contain procedures for evaluating your processes. In general these procedures come in three parts. The first part is an assessment and characterization of past performance. This will tell you where you have been, and will provide a benchmark for comparisons on where you might go in the future. Since this part of the evaluation does not attempt to characterize the predictability of the process, we cannot know if the past performance will be indicative of the future. But, things being what they are in this world, it is unlikely that your process will spontaneously do better in the future than it has done in the past. Therefore, the use of the past performance of your process as a baseline for comparisons with what might be is reasonable.

The second part of the evaluation procedures give an estimate of what might be accomplished by improving the process aim. Methods for doing this might include manual adjustment procedures, SPC, algorithmic SPC, Exponentially Weighted Moving Averages, Auto-Regressive Integrated Moving Averages, or automatic process controllers. The method is not defined. The objective of this part of the process evaluation is to determine the *potential savings* that can be realized by using any of these methods of improving the process aim.

The third part of the evaluation procedures will allow you to estimate the minimum savings that you can achieve by operating your process predictably. Predictable processes are operating, by definition, with minimum variation. This reduction in variation that occurs as you learn how to operate your process predictably will lower your Effective Cost of Production. The estimated savings from this lower Effective Cost will provide a reasonable basis for deciding if it is worth your time and effort to learn how to operate your process predictably.

Finally, after the evaluation of what can be accomplished by improving the process aim, and after the assessment of the potential savings from operating your process predictably, you will probably still have an Effective Cost of Production that is greater than 1.00. Since this Effective Cost of Production for a predictable process that is operated on-target characterizes the best that you can do with your current process, the difference between this value and 1.00 represents the savings that you might capture by reengineering the process.

Thus, based on the potential savings for the different types of process improvement, the process evaluation procedures given here will help you to choose between doing nothing to your process, improving the process aim, learning how to operate the process predictably, and completely reengineering the process.

Chapter Two

Characterizing the Past Performance of Your Process

"How much is it going to cost?" is a question that every manager learns to ask before going into production. And the answer to this question becomes the nominal cost of production. However, this nominal cost of production will typically not include those hidden costs associated with things that do not work as expected. Therefore, your actual cost of production will usually be somewhat higher than your nominal cost of production. And the ratio of the actual cost to the nominal cost will be your Effective Cost of Production. How do you determine the Effective Cost of Production for your own process? The following procedure will allow you to obtain a reasonable approximation for an existing process.

2.1 Choose Critical Characteristics

While any given product will generally have several characteristics specified, not all of these characteristics will be equally important to the customer. Moreover, among these many different characteristics, some will be more difficult for the manufacturer to produce within the specifications. For the purposes of characterizing the past performance of your production process we shall define the critical characteristics to be those characteristics that are important to your customer and which are difficult to make within the specifications. Given these two aspects of a critical characteristic, you will usually already know what your critical characteristics are for any given product simply because of the energy already expended in discussing this product with your customer.

For these critical characteristics you will need some data. These data should cover a reasonably extended period of time, and should all be collected following the last known process change. (You will want to use data that characterize your current process.) While it will be preferable to have 50 or more data, reasonable values may be obtained with as few as 20 data.

As an example throughout this chapter I will use the following data for the weights of 20 parts in grams:

717	719	718	717	718	715	716	719	721	719
717	711	711	706	708	710	708	713	712	710

2.2 Compute a Measure of Scale

In order to use the tables you will need to compute certain statistics. Among these values will be a *scale factor* for your data (also known as a measure of dispersion).

For the purpose of characterizing the past performance of your process this *scale factor* should be computed globally, using all of the data in a single computation. The preferred global measure of scale is the standard deviation statistic, *s*, where *s* is defined as:

$$s = \sqrt{\frac{\sum_{i=1}^{i=n}(X_i - \bar{X})^2}{n-1}}$$

The value of this statistic tells you how many measurement units are required to equal one standard unit of dispersion. It is easily obtained using handheld calculators or spreadsheet programs, and can be used as your *scale factor* for characterizing the past performance of your process.

For the Weight Data in the previous step:

scale factor = *s* = 4.46 *grams per standard deviation unit*

This scale factor is a conversion factor—it allows you to convert values from the measurement unit scale to a scale having standard units. This conversion makes it possible to look values up in tables instead of having to repeatedly perform complex computations.

An alternative way of finding a scale factor for your data would be to use the range of all the data. The range statistic is the difference between the maximum value and the minimum value. While the range of a large data set is less efficient (in a mathematical sense) than the standard deviation statistic, and while it is sensitive to extreme values, it is also much easier to explain to those who have little technical training than is the standard deviation statistic. So while the standard deviation statistic is the preferred global measure, the use of the range of all the data is a simple alternative which may be used when simplicity is more critical than efficiency.

To convert a range into a scale factor you will have to divide the global range by the appropriate constant, d_2. The value of d_2 is the average number of standard deviation units that a given global range will represent, and so the value of d_2 will depend upon the number of values in the original data set, n. Some values for d_2 are given in Table 2.1. A more complete set of values is given in Table 38.

Table 2.1: Constants for Use with Global Ranges

n	20	21	22	23	24	25	26	27	28	29	30	31
d_2	3.74	3.78	3.82	3.86	3.90	3.93	3.96	4.00	4.03	4.06	4.09	4.11
n	32	33	34	35	36	37	38	39	40	42	44	46
d_2	4.14	4.16	4.19	4.21	4.24	4.26	4.28	4.30	4.32	4.36	4.40	4.43
n	48	50	55	60	65	70	75	80	85	90	95	100
d_2	4.47	4.50	4.57	4.64	4.70	4.75	4.81	4.85	4.90	4.94	4.98	5.02

For the Weight Data the maximum value is 721 grams, while the minimum value is 706 grams, giving a range of 15 grams. This range is based on 20 values, so our scale factor will be:

$$\begin{aligned} \textit{scale factor} &= \frac{R}{d_2} = \frac{15 \textit{ grams}}{3.74 \textit{ standard deviations}} \\ &= 4.01 \textit{ grams per standard deviation} \end{aligned}$$

This value of 4.01 means that it takes about four measurement units to correspond to one standard unit of dispersion. Notice that this value is reasonably close to the value of the standard deviation statistic found earlier.

2.3 Characterize the Elbow Room

If your critical characteristic has two-sided specifications then the distance between the specifications can be considered to be the elbow room for your production process. (This distance is also known as the specified tolerance.) This specified tolerance can be converted into standard deviation units by dividing the distance between the specifications by the scale factor found in the previous step. Thus, our *Elbow Room* statistic will be:

$$Elbow\ Room = \frac{USL - LSL}{scale\ factor}$$

For the Weight Data the specifications are 710 ± 7 grams. Thus the specified tolerance would be 14 grams, which is converted into standard deviation units to be:

$$Elbow\ Room = \frac{14\ grams}{4.46\ grams\ per\ standard\ deviation} = 3.14\ std.\ dev.$$

This value means that the Elbow Room is equivalent to about 3.1 standard units of dispersion. Since processes will generally require a minimum of six standard deviations of Elbow Room, it is common to divide the value above by 6.0 to obtain the *performance ratio:*

$$P_p = \frac{Elbow\ Room}{6.0} = 0.52$$

which is interpreted to mean that the Elbow Room available was about 52 percent as wide as it would need to be (based on the past process outcomes). This performance ratio value is one of the numbers that you will need in order to find your Effective Cost of Production.

2.4 Characterize the Distance to Nearer Specification

For this step you will need to find the distance between the average of your data and the nearer specification. If we let *USL* represent the upper specification limit, and let *LSL* represent the lower specification limit, then the distance to the nearer specification, in measurement units, will be the smaller of two distances:

$$\textit{distance to upper specification} = DUS = USL - \bar{X}$$
$$\textit{distance to lower specification} = DLS = \bar{X} - LSL$$

The *Distance to Nearer Specification* statistic is simply the smaller of the two numbers above divided by your scale factor:

$$\textit{Distance to Nearer Specification} = DNS = \frac{\text{minimum of } DUS \text{ and } DLS}{\textit{scale factor}}$$

For the Weight Data the average of the 20 values is 714.25. The specifications for this part are 710 ± 7 grams. Thus we get:

$$DUS = USL - \bar{X} = 717 - 714.25 = 2.75 \textit{ grams}$$
$$DLS = \bar{X} - LSL = 714.25 - 707 = 7.25 \textit{ grams}$$

so that the Distance to Nearer Specification is:

$$DNS = \frac{2.75 \textit{ grams}}{4.46 \textit{ grams per standard deviation}} = 0.62 \textit{ standard deviations}$$

This value means that the average is about 0.62 standard deviation units away from the nearer specification limit. This number is commonly divided by 3.0 to obtain a *centered performance ratio*:

$$P_{pk} = \frac{DNS}{3.0} = \frac{0.62}{3.0} = 0.21$$

This centered performance ratio is the second value that you will need to determine your Effective Cost of Production.

2.5 Determine the Effective Cost of Production

Finally, in addition to your Elbow Room and *DNS* values (and their equivalent performance ratios), you will need to determine which table to use to find the Effective Cost of Production.

Tables 1 to 20 are for use with two-sided specifications. Table 1 is set up for the situation where all nonconforming product is scrapped. Table 2 is set up for the situation where all nonconforming product is reworked. Tables 3 to 20 are set up for rework on one side and scrap on the other side of the specifications.

Among Tables 3 to 20 the odd numbered tables cover the situation where the average is on the scrap side of the target value, while the even numbered tables cover the situation where the average is on the rework side of the target value.

The difference between Tables 3, 5, 7, etc. is the ratio of the cost of rework to the cost of scrap. Recall that the Curly Brackets expression was:

$$\textit{Curly Brackets} = \left\{ \textbf{ISP} + \textbf{IBT} + \frac{C_R}{C_S} \left[\textbf{IAT} + \textbf{IRP} \right] \right\}$$

Clearly, the value of Curly Brackets will depend upon the relationship between the cost of rework and the cost of scrap. Tables 3 and 4 are for the case where the cost of rework is equal to the cost of scrap. Tables 5 and 6 have this ratio set at 0.80. Tables 7 and 8 have this ratio equal to 0.67. And Tables 9 through 20, in pairs, have costs of rework set at 0.50, 0.33, 0.20. 0.10, 0.01, and 0.001, respectively. A complete guide to Tables 1 to 20 is provided on pages 97 and 98.

Each of these tables give the Effective Cost of Production for different combinations of C_p and C_{pk}. For the purpose of characterizing the past performance of your process, you will substitute the value of P_p for C_p, and P_{pk} for C_{pk}, and read out the Effective Cost of Production from the table.

The Effective Cost of Production combines the cost of scrap, the cost of rework, and the cost of variation about the target value into a single number. This number tells you how close you are to the nominal cost of production.

For example, a value of 1.28 would mean that your actual cost of producing an item is likely to be about 28 percent greater than the nominal cost of production. A value of 2.45 would mean that your actual cost should be about 2.45 times the nominal cost, meaning that your excess costs amount to 145 percent of the nominal cost.

The difference between the Effective Cost of Production and a value of 1.00 represents the excess costs associated with your production environment. These costs are cutting into your profits and eroding your position. Ways to reduce the Effective Cost of Production will be considered in Chapter Three.

Table 2.2: A Portion of Table 4: The Effective Cost of Production When the Cost of Rework is Equal to the Cost of Scrap and the Process Average is on the Rework Side of the Target

C_{pk}	C_p 0.10	0.20	0.30	0.40	0.50	0.60
0.60						1.28
0.50					1.37	1.29
0.40				1.50	1.38	1.34
0.30			1.68	1.49	1.43	1.42
0.20		1.95	1.65	1.53	1.50	1.51
0.10	2.36	1.86	2.66	1.60	1.59	1.61

For the process represented by the weight data, overweight parts are reworked while underweight parts are scrapped. The average of 714.5 grams is greater than the target value of 710 grams, thus the average is on the rework side of the target. We therefore would use Table 4 (given here, in part, as Table 2.2) to obtain an Effective Cost of Production:

P_p = 0.52 is rounded off to the closest value shown = 0.50,
P_{pk} = 0.21 is rounded off to the closest value shown = 0.20,
and so the Effective Cost of Production is approximately 1.50.

Thus, for the weight data, the producer should expect his actual costs of production to run about 50 percent higher than his planned costs of production. This excess cost of 50 percent over planned cost is money that he should not be spending.

The following chapters will outline ways to recapture most of these excess costs, and thereby to maximize the profitability of your current process.

2.6 Practice

- What would change about the Weight Data example if the global range was used to define the scale factor instead of the standard deviation statistic?

- The Length Data: the lengths of 20 parts, in units of a thousandth of an inch, are shown below. The 20 parts were obtained by selecting four parts from each of 5 successive shipments. The length is the critical dimension for this part.

shipment	-1-	-2-	-3-	-4-	-5-
	422	416	411	412	411
	424	416	413	411	413
	420	417	411	410	412
	420	413	408	414	410

The specifications for this dimension are 416 ± 7. Undersized parts are scrapped, while oversized parts are reworked. Since rework is essentially the same as the original processing, consider the cost of rework to be equal to the cost of scrap.

Use the following worksheet to evaluate the Effective Cost of Production for the product represented by these values.

Worksheet for Two-Sided Specifications

Part One: Characterizing Past Performance

1. Identify a critical product characteristic:

Product Identification: ________________________________

Critical Characteristic Evaluated: ________________________

Upper Specification Limit = *USL* = ____________

Target Value = *target* = ____________

Lower Specification Limit = *LSL* = ____________

2. Compute a *scale factor* using a global measure of dispersion:

number of values used = N = ______

scale factor: s = ____________ or R/d_2 = ____________

3. Characterize the *Elbow Room*:

specified tolerance = $USL - LSL$ = ____________

Divide by your *scale factor* to obtain the

Elbow Room = ―――――――― = ____________

Divide by 6.0 to obtain your

Performance Ratio, P_p = [ ]

4. Characterize the *Distance to Nearer Specification*:

average value for the N data = ____________

$USL - average = DUS$ = ____________

$average - LSL = DLS$ = ____________

$$DNS = \frac{\text{smaller of } DUS \text{ and } DLS}{scale\ factor} = \text{――――――} = \text{____________}$$

Divide the *DNS* value by 3.0 to obtain your

Centered Performance Ratio, P_{pk} = [ ]

Worksheet for Two-Sided Specifications

page 2 of 4

5. Find the Effective Cost of Production: Fill in the blanks below:

Average = ________

scrap? *scrap?*
rework? *LSL* = _______ *target* = ________ *USL* = _______ *rework?*

What happens to product that exceeds the upper specification?

What happens to product that falls below the lower specification?

What is the cost of rework as a proportion of the cost of scrap?

100% 80% 67% 50% 33% 20% 10% 1% 0.1%

Is the average on the rework side or the scrap side of the target?

Which Table fits your situation? Table _____

From this table, with P_p = , and with P_{pk} =

read off your **Effective Cost of Production**: [ ]

The amount by which this value exceeds 1.00 represents the excess costs associated with your past production.

(The second part of this worksheet will be given in the next chapter.)

The appendix contains a copy of all of the worksheets, which you may copy for your own use in performing process evaluations.

- For the Length Data your scale factor is either $s = 4.40$ or $R/d_2 = 4.28$.
- Using the scale factor of 4.40 you get an Elbow Room of 3.18 std. dev., a P_p of 0.53, a *DNS* of 1.18 std. dev., and a P_{pk} of 0.39.
- Using the scale factor of 4.28 you get an Elbow Room of 3.27 std. dev., a P_p of 0.54, a *DNS* of 1.21 std. dev., and a P_{pk} of 0.40.
- Using Table 3, rounding P_p to 0.5 and P_{pk} to 0.4 you find an Effective Cost of Production of 1.41, 41% greater than the nominal cost.

Chapter Three

The Payoff for Process Improvement

The Effective Cost of Production will characterize the past performance of your process. The difference between your Effective Cost of Production and the optimum value of 1.00 is the potential savings on this characteristic of this product. The greater this difference, the greater your potential savings. If the potential savings are not great enough to justify the cost of making process improvements then you should look for greater opportunities elsewhere.

3.1 Operating On-Target

If the average value for the past performance data is not close to the target value then there will be losses associated with being off-target. To see what you can save by operating your process on-target you can recompute the Effective Cost of Production with the centered performance ratio, P_{pk}, equal to the performance ratio, P_p:

To find the *Centered Cost of Production*,
change the value of P_{pk} to be the same as the value of P_p
and using the same table as before,
find a new value for the Effective Cost of Production.

For the Weight Data the P_p value was 0.52 while the P_{pk} value was 0.21, and the Effective Cost of Production was 1.50. The *Centered Cost of Production* is found by letting both performance ratios be equal to 0.50. Using Table 4 again (part of which is reproduced as Table 3.1) we get:

$$\textit{Centered Cost of Production} = 1.37$$

This value of 1.37 amounts to a potential savings of 13 percent of the nominal cost. This is the amount that can be realized by operating this process with an average near 710 grams.

Table 3.1: A Portion of Table 4

	C_p *values*					
C_{pk}	***0.10***	***0.20***	***0.30***	***0.40***	***0.50***	***0.60***
0.60	-	-	-	-	-	1.28
0.50	-	-	-	-	1.37	1.29
0.40	-	-	-	1.50	1.38	1.34
0.30	-	-	1.68	1.49	1.43	1.42
0.20	-	1.95	1.65	1.53	1.50	1.51
0.10	2.36	1.86	2.66	1.60	1.59	1.61

Of course, a process does not spontaneously operate on-target. Nor does a process remain fixed and unwavering over an extended period of time. To operate your process on-target will require that you have some feedback mechanism that will let you know when your process is on-target and when it is off-target. In addition you will need to have an appropriate adjustment mechanism. And then you will need to train the process operators on how to use the feedback to make appropriate adjustments. When you do all this, and maintain this mode of operation, then you will be able to save the difference between the Effective Cost of Production and the Centered Cost of Production.

However, the critical phrase in the previous paragraph is "appropriate adjustments." If the adjustments are inappropriate they will invariably increase the variation; which will reduce the P_p value. A glance at Table 3.1 will show what effect this would have on your cost of production. At the same time that you would be trying to move up in the table by improving your aim, the increased variation would force you to the left. Instead of moving from 1.50 to 1.37, your costs could stay the same, or increase to 1.53, or 1.65, or even 1.95.

So it is not just a matter of telling the operators to keep the process on-target, but rather having an adjustment procedure that matches the action with the information coming from the process. One of the simplest manual techniques to match up the actions with the information is to use a process behavior chart as a definition of when to make a process adjustment as well as a way of evaluating each adjustment.

An alternative to manual process adjustment would be an automated process controller. However, to realize the potential savings of operating on-target the automated process controller will also have to make appropriate adjustments. Just as with a manual system, inappropriate adjustments will invariably make things worse. Anything which increases the overall variation of the process outcomes will decrease the P_p value, and inspection of the values in Table 3.1 will show that this will increase the Effective Cost of Production, no matter how much we improve the process aim.

Finally, the difference between the Effective Cost of Production and the Centered Cost of Production is about the *most* that you can expect to save with a process adjustment approach. To achieve greater savings will require a different approach.

3.2 Operating Your Process Predictably

Process adjustments made in response to process changes will always be reactive in nature. Such an operating philosophy can not anticipate process changes. Moreover, since a process adjustment approach does not seek to eliminate the causes of the process changes, but merely to compensate for the changes in a timely manner, a process adjustment approach is more of a maintenance technique than an improvement. This is why there is a limited amount of savings that can be obtained from a process adjustment approach.

A more proactive approach is to seek to operate your process predictably. The distinction between a predictable process and an unpredictable process can be traced back to the work of Walter Shewhart, and the way that we characterize a process as being one or the other is by the use of a process behavior chart.

A process is said to be operated predictably when, through the use of past experience, we can describe, at least within limits, how the process will behave in the future. If your process has behaved consistently in the past, then it is reasonable to assume that it will continue to behave consistently in the future. However, if your process has behaved inconsistently in the past, then how can you expect it to do otherwise in the future? Therefore the process behavior chart examines the past behavior of the data to see which type of variation is present.

When a process is predictable it will display routine variation, which can be thought of as the result of many different cause-and-effect relationships, none of which have a dominant influence on the process.

When a process is unpredictable it will display exceptional variation in addition to the routine variation. This exceptional variation is said to be due to assignable causes which, by definition, dominate the common causes of routine variation.

Thus, Shewhart's distinction is primarily a way to determine how far we have come in our understanding of our process. When we design a product or a process we will usually begin with a mental list of the known cause-and-effect relationships. This list can have dozens, or even hundreds of causes. Fortunately for us there is usually a Pareto principle at work—a few of these causes will have dominant impacts upon our process outcomes while the remainder will have much smaller effects. Since we can never exert active control over all of these cause-and-effect relationships, we try to control those causes which have the dominant effects, and allow the other causes to vary without making any effort to control them.

But the need to get into production will often cut short the development time, forcing R&D to turn the process over to manufacturing before all of the dominant cause-and-effect relationships have been identified. Thus, at start-up, most processes operate erratically and inconsistently. Band-aids and quick fixes are applied, and the process is considered to be up and running. However, there may still be dominant cause-and-effect relationships which are not being controlled. And when dominant factors are overlooked the process will continue to behave erratically in spite of all your quick fixes. Moreover, as wear and tear increase, the process will tend to become more sensitive to other cause-and-effect relationships—factors that formerly had a small impact may begin to have a large impact. And the result will be an unpredictable process—one that is subject to dominant cause-and-effect relationships (which are known as assignable causes).

The process behavior chart allows you to detect the presence of these assignable causes as they occur, when they occur, so that you can take action to remove their effects from your process. This type of action is proactive, rather than reactive—instead of compensating for a past upset, you are modifying the process to prevent further upsets. And, as a side effect, you will find that by identifying assignable causes and removing their effects from your process you will also be improving your process outcomes and reducing your Effective Cost of Production.

3.3 The Payoff for Operating Predictably

The process behavior chart serves as the operational definition of when a process is being operated predictably. It defines the maximum degree of consistency that your process is capable of achieving; it provides a mechanism for moving your process toward maximum consistency; and it provides a basis for judging how close you have come to achieving that maximum consistency.

If a process is predictable, then it is essentially being operated with maximum consistency and the natural process limits will define the *actual capability* of the process.

More commonly a process will be unpredictable. An unpredictable process is not being operated as consistently as it could be. While the width of the natural process limits will define the *hypothetical capability* for such a process, the fact that the process is subject to assignable causes means that the process will not achieve this hypothetical capability until something is done about removing the effects of the assignable causes.

Therefore, the first step in determining the payoff for predictability will be to place your process data on a process behavior chart. If your data were collected over an extended period of time, and if they do not display evidence of unpredictability, then you may have a predictable process. In this case the *Centered Cost of Production* computed earlier is likely to be the best your current process can do.

If your process data display evidence of unpredictability, then you can use the process behavior chart as the basis for determining how much better your process can be if you make the effort to operate your process predictably.

The Weight Data are shown on an *XmR* chart in Figure 3.1. There are points outside the limits and long runs above and below the central line.

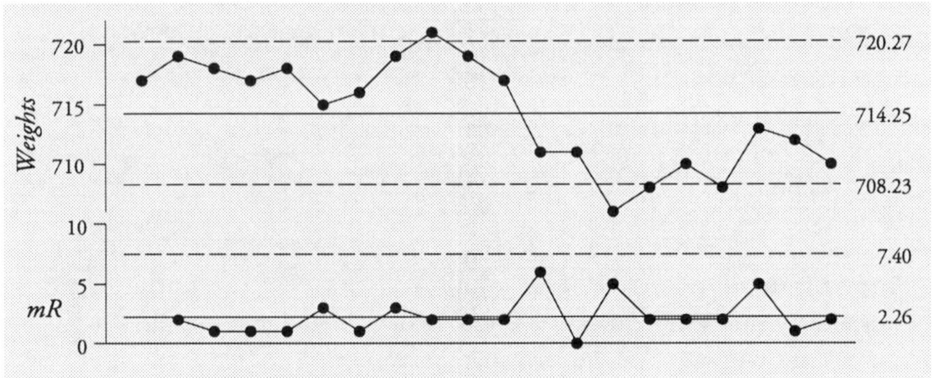

Figure 3.1: The *XmR* Chart for the Weight Data

Figure 3.1 shows a process that is not being operated predictably. It could be operated more consistently. The following operations will allow you to approximate just how much more consistently this process could be operated.

To estimate the scale factor that would apply when this process is operated with maximum consistency we will use the average moving range, which in this case is 2.26 grams. Since the moving ranges are based on successive differences, they have an effective subgroup size of $n = 2$. From Table 38 we find that the d_2 value for subgroups of size 2 is 1.128 standard deviations. Thus we estimate a *scale factor* for predictable operation to be:

$$\textit{predicable scale factor} = \frac{\bar{R}}{d_2} = \frac{2.26 \textit{ grams}}{1.128 \textit{ std. dev.}} = 2.00 \textit{ grams per std. dev.}$$

Using this new scale factor we can recharacterize the Elbow Room. While the specifications are still 710 ± 7 grams, the Elbow Room is now:

$$\textit{Elbow Room} = \frac{14 \textit{ grams}}{2.00 \textit{ grams per std. dev.}} = 7.0 \textit{ std. dev.}$$

Which, when divided by 6.0, gives a *hypothetical capability ratio*:

$$C_p = \frac{7.0}{6.0} = 1.17$$

While you could also reevaluate the Distance to Nearer Specification using the original average from your data, it is more likely that as you learn how to operate your process more predictably you will also be able to operate closer to the target value. Therefore, it is reasonable to assess the payback from operating your process predictably and on-target by assuming that C_{pk} is equal to the hypothetical capability ratio above.

Table 3.2: A Portion of Table 4

	C_p *values*									
C_{pk}	***0.30***	***0.40***	***0.50***	***0.60***	***0.70***	***0.80***	***0.90***	***1.00***	***1.10***	***1.20***
1.20	-	-	-	-	-	-	-	-	-	1.08
1.10	-	-	-	-	-	-	-	-	1.09	1.08
1.00	-	-	-	-	-	-	-	1.11	1.10	1.10
0.90	-	-	-	-	-	-	1.14	1.12	1.12	1.14
0.80	-	-	-	-	-	1.17	1.15	1.15	1.16	1.19
0.70	-	-	-	-	1.22	1.18	1.18	1.20	1.22	1.25
0.60	-	-	-	1.28	1.23	1.22	1.24	1.26	1.29	1.32
0.50	-	-	1.37	1.29	1.28	1.29	1.31	1.34	1.37	1.40
0.40	-	1.50	1.38	1.34	1.35	1.37	1.40	1.43	1.46	1.49
0.30	1.68	1.49	1.43	1.42	1.44	1.46	1.49	1.52	1.55	1.58
0.20	1.65	1.53	1.50	1.51	1.54	1.57	1.59	1.62	1.64	1.66
0.10	2.66	1.60	1.59	1.61	1.64	1.66	1.69	1.71	1.73	1.75

To use Table 3.2 we round both C_p and C_{pk} from 1.17 to 1.20 and find the corresponding table entry:

Potential Cost of Production = 1.08.

Thus, by learning how to operate this process predictably and on-target you can reduce your Effective Cost of Production from 150 percent of the nominal cost to 108 percent of the nominal cost—a savings of 42 percent of the nominal cost on each unit produced. Of course, these savings remain hypothetical until you actually take action to operate this process predictably. But, before you have made the effort, you have a value for the potential savings that can be achieved. This 42 percent reduction in the cost of production is what you can do with your *current* process. No capital expense is required. No new technology is required. No expensive process upgrade is needed. Simply by learning how to get the most out of your current process you can reduce your Effective Cost of Production from 1.50 to 1.08.

Moreover, since the computation of the Potential Effective Cost of Production is based on the data obtained while the process was unpredictable, you may actually do even better than the value found. The 42 percent reduction is the *minimum* you should be able to achieve by operating your process predictably and on-target.

Thus, by using a scale factor based on a process behavior chart you can obtain a Potential Cost of Production. By comparing this value with the Effective Cost of Production you can evaluate the payback that can be had from learning how to operate your process predictably and on-target.

3.4 The Reengineering Dividend

If you are thinking of reengineering your process, or performing a process upgrade, or doing some experiments to try to optimize your process, how should you evaluate your potential payoff? Which number should you use in a cost/benefits analysis? Do you use the Effective Cost of Production for the past performance of your current process? Or do you use the Centered Cost of Production value? Or do you use the Potential Cost of Production for a predictable process that is operated on-target?

Since the last of these three numbers defines what you can do with your current process, it is the number that you should use to evaluate the need for Reengineering your process. Only when the Potential Effective Cost of Production is substantially greater than 1.00 will the payoff of reengineering the process offset the cost of the reengineering effort.

For the Weight Data the Effective Cost of Production was 1.50, the Centered Cost of Production was 1.37, and the Potential Cost of Production was 1.08. Therefore the maximum that you can hope to save by further efforts at improving the current process is 8 percent of the nominal cost of production. So unless you can substantially change the nominal cost of production itself, further improvement efforts may cost more than they are worth.

You may think that if you leap from your current unpredictable process to a new and improved process that you should use the Effective Cost of Production (based on past performance) as the basis for evaluating the payback. While this seems plausible, the problem comes in the execution of the new and improved process. If you are not operating your current process predictably, then what makes you think that you will operate your new process predictably? Experience shows that if you cannot get the most out of your current process, you will not get the most out of a new and improved process. Therefore, while the Potential Cost of Production may be hypothetical, it is achievable with your current technology and your current workforce.

Operating a process predictably and on-target is equivalent to getting the process to operate up to its full potential. If you have the discipline to do this, then you may find that, in most cases, further improvement efforts are not needed. If you do not have the discipline to operate your processes predictably and on-target, then other improvement efforts are doomed to be less effective than they could be. Think it over. Operating predictably and on-target is an offer you cannot afford to refuse.

3.5 Practice

- The lengths of 20 parts, in units of a thousandth of an inch, are shown on the next page. The 20 parts were obtained by selecting four parts from each of 5 successive shipments. The length is the critical dimension for this part. The specifications for this dimension are 416 ± 7.

Undersized parts are scrapped, while oversized parts are reworked.

Since rework is essentially the same as the original processing, consider the cost of rework to be equal to the cost of scrap.

The formulas and constants for constructing Average and Range charts are given in Table 40. To use these formulas and constants you will need to compute the Grand Average and the Average Range.

Grand Average = $\bar{\bar{X}}$ = ____________

Average Range = $\bar{R}$ = ____________

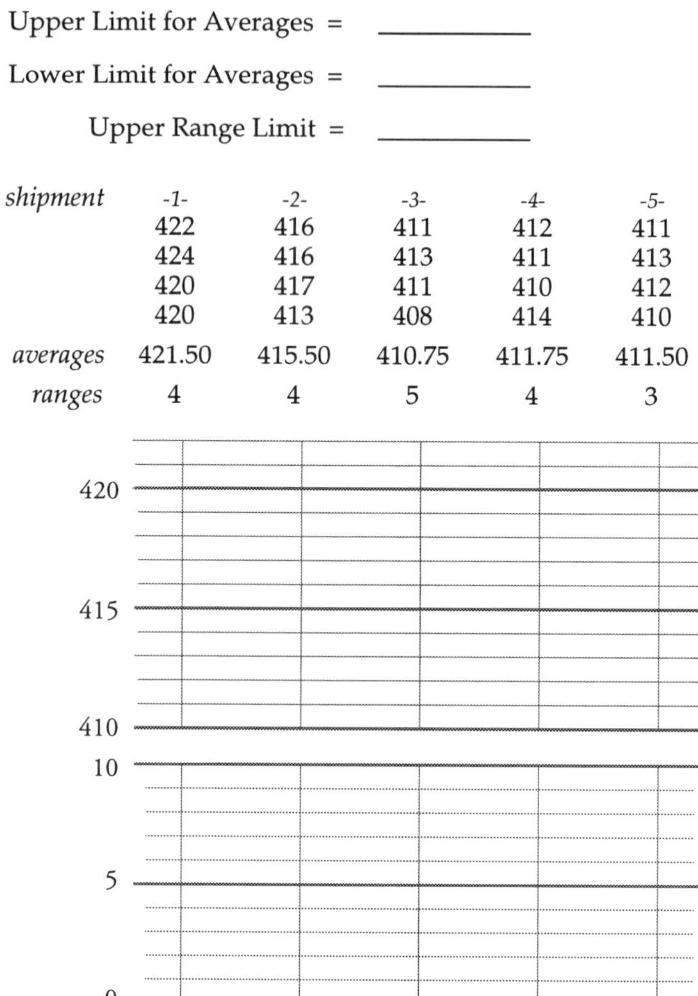

Upper Limit for Averages = ___________

Lower Limit for Averages = ___________

Upper Range Limit = ___________

shipment	-1-	-2-	-3-	-4-	-5-
	422	416	411	412	411
	424	416	413	411	413
	420	417	411	410	412
	420	413	408	414	410
averages	421.50	415.50	410.75	411.75	411.50
ranges	4	4	5	4	3

• Use the following worksheets to assess the paybacks for various improvement strategies for this process.

Worksheet for Two-Sided Specifications

Part Two: Potential Payoffs for Improvements

6. Find your *Centered Cost of Production*:

Using the table identified on page 22, and with $P_{pk} = P_p$ find the Effective Cost of Production for a centered process:

Centered Cost of Production = ____________

Subtract the Centered Cost of Production from the Effective Cost of Production to find the

Potential Benefit of Operating On-Target = [ ]

(Multiply by nominal cost of production to get benefit in dollars.)

7. Place your Past Performance Data on a Process Behavior Chart:

☐ If your process appears to be predictable skip Steps 7, 8, and 9. Unless or until your process is changed in some major way, the cost of production in Step 5 is what you should expect from your process in the future. If you improve your process aim you can expect to get the cost of production found in Step 6.

☐ If your process appears to be unpredictable, then you will need to compute a *Potential Cost of Production*. As a first step toward this value use the Average Range, or the Average Moving Range, from your process behavior chart and the appropriate bias correction factor from Table 38 to compute a *predictable scale factor:*

subgroup size, n = _____ (use $n = 2$ for *XmR* chart)

Average Range = $\bar{R}$ = __________

$$\textit{predictable scale factor} = \frac{\bar{R}}{d_2} = \_\_\_\_\_\_\_\_ = \_\_\_\_\_\_\_\_$$

Worksheet for Two-Sided Specifications

page 4 of 4

8. Find your hypothetical process capability:

Divide your specified tolerance in measurement units

by your *predictable scale factor* = ______________ = ______________

Divide this value by 6.0 to obtain your

Hypothetical Capability Ratio, C_p = ______________

9. Find the Potential Cost of Production:

Assume that C_{pk} is equal to C_p

and use the table identified in Step 5

to find the ***Potential Cost of Production***: [ ]

Subtract the Potential Cost of Production

from the Effective Cost of Production found in Step 5

to obtain the minimum potential benefit

to be had by operating predictably and on-target:

Minimum Potential Benefit from Predictable and On-target [ ]

Multiply by the nominal cost of production to turn this into dollars.

Chapter Four

One-Sided Product Specifications

Process evaluation for a characteristic having a one-sided specification differs from that for a characteristic having two-sided specifications. In particular, the concept of Elbow Room is no longer well defined and the definition of the target value changes. Therefore different tables, different procedures, and different worksheets are needed.

4.1 Different Types of One-Sided Specifications

Some one-sided specifications are minimums or maximums that provide a safety threshold. For example, the government has set a minimum tensile strength for the rayon cord used in automobile brake hoses. In order to have a safety margin, automobile manufacturers have their own, even higher, minimum tensile strength, and the automotive suppliers use an even higher minimum standard.

There is a standard on the maximum concentration of hydrogen gas that can be allowed in the airspace within barrels shipped by ground transport. In order to have a safety margin, chemical companies set their own internal specifications on the concentration levels of a given compounds.

With one-sided specifications like those above you do not want to operate anywhere near the specification limit. You will commonly take action long before you reach a safety specification. Therefore, while such specifications may define the point where some regulatory agency can take legal action, they do not define the point were you will want to take action on the product. Therefore, while *safety specifications* are specifications, they are not *product specifications*.

Then there are the one-sided specifications known as net weights. In filling containers you will need to have some minimum percentage with fill weights that meet or exceed the net weight on the label. Yet you do not screen the filled containers to remove the underweight containers. If you do not have the minimum percentage required, then you take action to increase the average fill weight. Even though the specification applies to the individual items, the action is not taken on the individual items, but rather on the process itself. This makes net weight specifications into *process specifications* rather than *product specifications* in spite of the fact that they specify a product characteristic. Whenever a specification includes a conformance percentage it becomes a process specification rather than an product specification. Process specifications will be covered in Chapter Seven.

The specifications of interest in this chapter are one-sided *product specifications*—specifications where action is taken to scrap or rework the items whenever they exceed some minimum or maximum value. Inherent in having a point for taking action is the concept of cutting your losses, and this, in turn is based on the concept of decreasing functionality as the item approaches the action point. Therefore, to realistically deal with one-sided product specifications we will need to evaluate both the excess costs associated with nonconforming items and the excess costs associated with decreased functionality on the conforming side of the limit. The excess cost function that will do this will have the form shown in Figure 4.1.

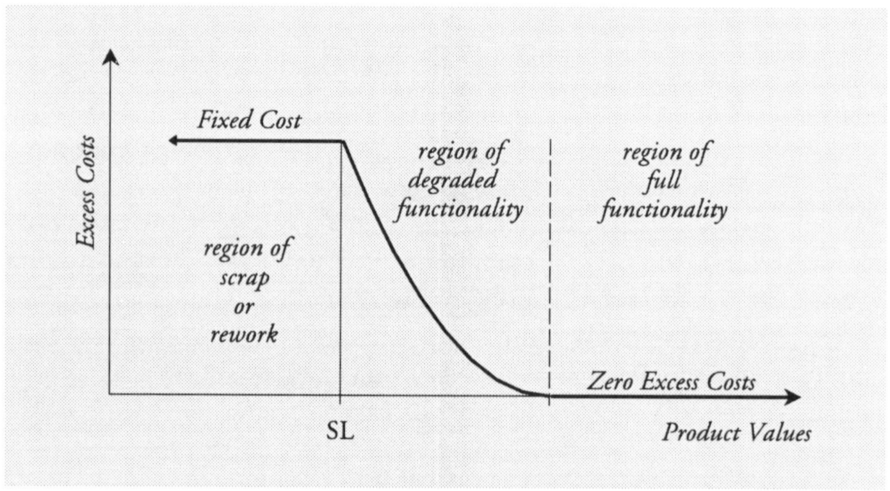

Figure 4.1: Excess Cost Function for a One-Sided Product Specification

4.2 One-Sided Product Specifications Without Target

Unlike two-sided specifications, one-sided product specifications will seldom have a clear-cut target value. When this happens the problem of obtaining a realistic excess cost function is one of determining the point where the excess costs become zero. As before, the costs associated with nonconforming product will be fixed and fairly easy to identify. But where does the degradation of functionality, which eventually leads to the nonconforming product, begin? In terms of Figure 4.1, where do we place the dashed line?

When there is no explicit target value we will have to *choose* where to place the dashed line. By analogy with the two-sided specification case, it seems reasonable to place the zero excess cost point at the average value of a process having $C_{pk} = 1.0$. Therefore, the dashed line in Figure 4.1 is placed three standard deviations away from the one-sided specification, as shown in Figure 4.2.

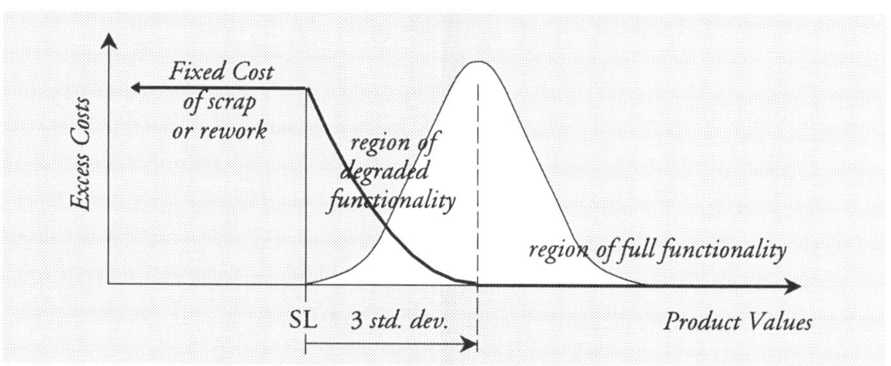

Figure 4.2: A Generic Excess Cost Function for a One-Sided Product Specification Without Explicit Target

Of course, if you have specific information about the region of degraded functionality, then you can turn this information into an explicit target and you will not need to use the generic target of Figure 4.2. Figure 4.2 is only intended to provide a way to deal with those situations without an explicit target. The excess cost function of Figure 4.2 was used to generate Table 21.

A portion of Table 21 is shown in Table 4.1. The first column of Table 21 is for the case where all nonconforming items are scrapped. These values were computed under the assumption that the cost of scrap is equal to the nominal cost of production. Columns Two through Eight are for the case where all nonconforming items are reworked. These seven columns use different ratios for the cost of rework versus the nominal cost of production. For any given column, the only value needed to obtain an Effective Cost of Production is the centered capability ratio, C_{pk}, or its past performance equivalent, the centered performance ratio, P_{pk}.

Table 4.1: A Portion of Table 21

	Scrap	*Rework (cost of rework / nominal cost)*						
C_{pk}		***1.00***	***0.80***	***0.67***	***0.50***	***0.33***	***0.20***	***0.10***
0.00	2.58	1.789	1.631	1.526	1.395	1.263	1.158	1.079
0.10	2.15	1.709	1.568	1.473	1.355	1.236	1.142	1.071
0.20	1.853	1.619	1.495	1.413	1.310	1.206	1.124	1.062
0.30	1.641	1.523	1.419	1.349	1.262	1.174	1.105	1.052
0.40	1.483	1.427	1.342	1.285	1.214	1.142	1.085	1.043
0.50	1.361	1.336	1.269	1.224	1.168	1.112	1.067	1.034
0.60	1.265	1.255	1.204	1.170	1.128	1.085	1.051	1.026
0.70	1.189	1.186	1.149	1.124	1.093	1.062	1.037	1.019
0.80	1.131	1.130	1.104	1.087	1.065	1.043	1.026	1.013
0.90	1.087	1.087	1.069	1.058	1.043	1.029	1.017	1.009
1.00	1.055	1.055	1.044	1.037	1.028	1.018	1.011	1.006
1.10	1.033	1.033	1.027	1.022	1.017	1.011	1.007	1.003
1.20	1.019	1.019	1.015	1.013	1.010	1.006	1.004	1.002
1.30	1.010	1.010	1.008	1.007	1.005	1.003	1.002	1.001
1.40	1.005	1.005	1.004	1.004	1.003	1.002	1.001	1.001
1.50	1.003	1.003	1.002	1.002	1.001	1.001	1.001	1.000
1.60	1.001	1.001	1.001	1.001	1.001	1.000	1.000	1.000
1.70	1.000	1.000	1.000	1.000	1.000	1.000	1.000	1.000

For example, consider the situation shown in Figure 4.2, where the process average is essentially three standard deviations away from a minimum specification. When *DNS* = 3 standard deviations the C_{pk} value will be 1.00. If the nonconforming products are scrapped, then the Effective Cost of Production will be 1.055. If we changed this process so that is average was 4.5

standard deviations away from the specification, then our C_{pk} would become 1.50, and our Effective Cost of Production would drop to 1.003.

In order to escape all excess costs and have an Effective Cost of Production of 1.000 we would need to shift the C_{pk} value to 1.7, which is equivalent to a process average that is 5.1 standard deviations away from the specification. This process average would avoid *all* of the excess costs, both those of nonconforming product and those associated with conforming, but degraded, product.

4.3 Optimum Aim Point for One-Sided Specifications

Of course, if there is no penalty for operating far from a one-sided specification, then it is best to do so. And a process average that is 5.1 standard deviations away from a one-sided specification would be reasonable. But what happens when there is a penalty for operating far from a one-sided specification? We can incorporate such a penalty into our calculations by using different nominal costs of production at different values for the process average—in effect, we can use a nominal cost function.

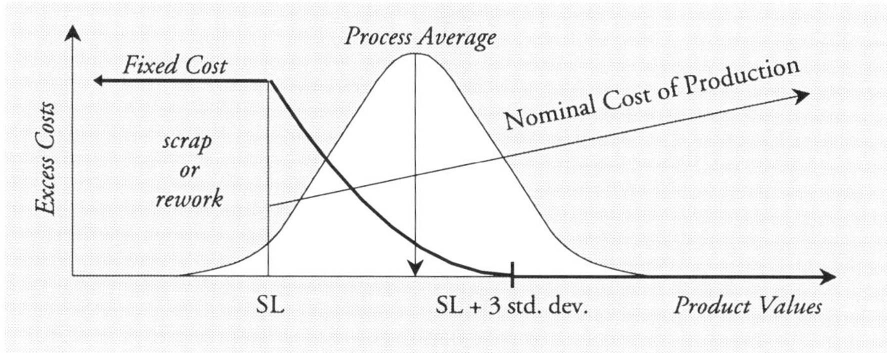

Figure 4.3: A Nominal Cost Function

Clearly, in Figure 4.3, the excess cost function will create a pressure to increase the process average. At the same time, the nominal cost function will create an opposite pressure to decrease the process average. We combine these two competing pressures when we consider the actual costs of production:

$$\textit{actual cost} = \textit{nominal cost} \times \textit{effective cost}$$

By multiplying the Effective Cost of Production for a given average by the nominal cost of production for that average you can obtain the actual cost of production for that average. By comparing several such different actual costs you can determine the appropriate optimum aim point for the process average for a given process.

Say that your product has a critical characteristic with a minimum specification. Assume that nonconforming product is scrapped. This means that you can use the first column of Table 21 to obtain values for the Effective Cost of Production (ECP). Next you will need to know the nominal costs of production (NCP) at each of the points defined by the C_{pk} values of Table 21. For example, say that when the process average is equal to the minimum specification the nominal cost of production is exactly \$2.00. At a C_{pk} of 0.00 the Effective Cost of Production is 258%. Therefore the actual cost of production at this aim point is:

$$\textit{actual cost} = \$\,2.00 \times 2.58 = \$5.16$$

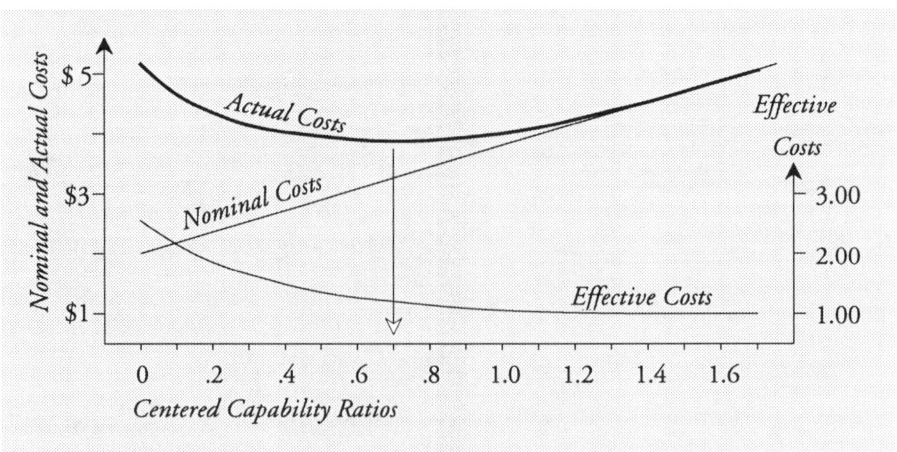

Figure 4.4: Effective Costs times Nominal Costs = Actual Costs

By repeating this computation at each of the C_{pk} values in Table 21, you can identify that aim point with the minimum actual cost. In this example the nominal cost is assumed to increase in equal increments as you move away from the minimum specification.

Table 4.2: Actual Costs for Different Process Aim Points

C_{pk}	*0.0*	*0.1*	*0.2*	*0.3*	*0.4*	*0.5*
ECP	2.58	2.15	1.853	1.641	1.483	1.361
NCP $	2.00	2.18	2.36	2.54	2.72	2.90
cost $	*5.16*	*4.69*	*4.37*	*4.17*	*4.03*	*3.95*
C_{pk}	*0.6*	*0.7*	*0.8*	*0.9*	*1.0*	*1.1*
ECP	1.265	1.189	1.131	1.067	1.055	1.033
NCP $	3.08	3.26	3.44	3.62	3.80	3.98
cost $	*3.90*	*3.88*	*3.89*	*3.94*	*4.01*	*4.11*
C_{pk}	*1.2*	*1.3*	*1.4*	*1.5*	*1.6*	*1.7*
ECP	1.019	1.010	1.005	1.003	1.001	1.000
NCP $	4.16	4.34	4.52	4.70	4.88	5.06
cost $	*4.24*	*4.38*	*4.54*	*4.71*	*4.89*	*5.06*

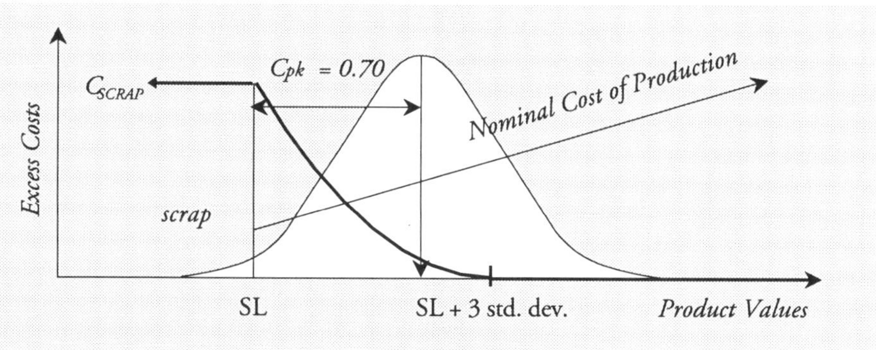

Figure 4.5: Optimal Aim Point for Table 4.2

Inspection of the Actual Costs of Production in Table 4.2 shows that the minimum cost per unit will occur when the centered capability ratio is near 0.70, which corresponds to a process average that is 2.1 standard deviations away from the specification. The computations in Table 4.2 combine the penalty for not operating close to the specification with the losses from scrap and the excess costs associated with variation about the process average. They allow you to determine the optimum aim point to use when a predictable process has a one-sided specification without a target.

Of course, if you do not operate your process predictably, it will be hard to maintain your process at a given aim point or to maintain a given amount of process variation. With an unpredictable process your actual cost of production will exceed the one computed in Table 4.2.

4.4 Characterizing Past Performance

For one-sided product specifications you can characterize past performance in the following manner:

Step One: Identify the critical characteristic and its one-sided product specification. This step is the same as it was with two-sided specifications.

Step Two: Compute a scale factor using a global measure of dispersion. In doing this you will need to have at least 20 data collected over an extended period of time. Fifty or more data are preferred.

Step Three: Characterize the Distance to Nearer Specification. (The word "nearer" is redundant, but it is used here to avoid introducing a new term for what is essentially the same computation performed earlier.) Obtain the distance between the average of your data and the one-sided product specification. Divide this distance by the scale factor to obtain the Distance to Nearer Specification:

$$DNS = \frac{|\,average - specification\,|}{scale\ factor}$$

And then divide this value by 3.0 to obtain your centered performance ratio:

$$P_{pk} = \frac{DNS}{3.0}$$

Step Four: Use Table 21 to determine the Effective Cost of Production. If nonconforming product is scrapped use column one. If nonconforming product is reworked, then determine the ratio of the cost of rework to the nominal cost of production. In doing this use the nominal cost of production at the average value. Then use the appropriate column from Table 21, along with the P_{pk} value, to find the Effective Cost of Production.

As an example, consider the following 20 data on the level of a certain contaminant in a given compound. The maximum allowable level is 150 ppm. The past 20 shipments have had the following levels of this contaminant:

143	144	116	117	124	125	124	117	110	106
104	110	110	123	125	127	109	98	91	92

The average of these 20 values is 115.75. The range is 53 ppm. Thus a scale factor for these data would be:

$$\textit{scale factor} = \frac{R}{d_2} = \frac{53\ ppm}{3.74\ std.\ dev.} = 14.1\ \textit{ppm per standard deviation}$$

The Distance to Nearer Specification is:

$$DNS = \frac{150\ ppm - 115.75\ ppm}{14.1\ ppm\ per\ std.\ dev.} = 2.43\ \textit{standard deviations}$$

Therefore the centered performance ratio is:

$$P_{pk} = \frac{DNS}{3.0} = 0.81$$

Using Table 21, and assuming that nonconforming product is recycled through the process, resulting in a rework cost that is essentially equal to the cost of production, and rounding the 0.81 to 0.80, we find the Effective Cost of Production to be 1.130. With a nominal cost of production of 32.4 cents per pound, this Effective Cost of Production translates into an actual cost of 36.6 cents per pound.

4.5 Assessing Process Aim Strategies

With a one-sided product specification and an incentive to operate close to that specification there is a need to determine the optimum aim point for your process. As was illustrated earlier, the Effective Costs of Production can be multiplied by the nominal costs to obtain the actual costs of production at different process aim points. The aim point with the smallest actual cost of production will be the optimum point for a process aim strategy.

What will a process aim strategy accomplish? When you have characterized past performance you can use the actual cost for the past performance as a point of comparison. The difference between the actual cost of past performance and the actual cost at optimum aim will generally be the *maximum* you can hope to save using a process aim strategy.

For the Contaminant Level Data, the historical average was 115.75 ppm, and the scale factor was 14.1 ppm per standard deviation. Since we are going to use the C_{pk} values in Table 21 in assessing the potential improvements we will need to convert them into levels of the contaminant in parts per million using the following expression:

$$\textit{aim point} = \textit{specification} \pm [\, C_{pk} \times \textit{scale factor} \times 3.0 \,]$$

where the plus is used with minimum specifications and the minus is used with maximum specifications. Thus, for this example:

$$\textit{aim point in ppm} = \textit{150 ppm} - [\, C_{pk} \times \textit{14.1 ppm per std. dev.} \times 3.0 \,]$$

Using this formula the C_{pk} values in Table 4.1 result in the aim points shown in Table 4.3.

Next we will need to obtain the nominal costs of production at each of the different aim points computed in Table 4.3. Say that you know, from past experience, that the energy costs for producing this compound, in dollars per pound, can be written as:

$$\textit{cost per pound in dollars} = \frac{37.5}{\textit{average level of contaminant in ppm}}$$

Using this formula we get the nominal costs of production shown.

Finally, since nonconforming product is recycled at 100% of the nominal cost we use the second column of Table 21 to obtain the Effective Costs of Production for the 18 aim points considered in Table 4.3. These are multiplied times the nominal costs to get the actual costs shown.

Comparing the minimum actual cost shown in Table 4.3 with the actual cost found by characterizing past performance, we can assess the maximum potential to be achieved by improving the process aim. The minimum cost in Table 4.3 is 36.4 cents per pound, which occurs at 112 ppm. In the past this process averaged 116 ppm at an average cost of 36.6 cents per pound.

About the most that we can expect to gain from a strategy for improving the process aim is 0.2 cents per pound, which at 25 million pounds per year amounts to $ 50,000 annually.

Table 4.3: Finding the Optimum Aim for the Contaminant Level Data

C_{pk}	***0.0***	***0.1***	***0.2***	***0.3***	***0.4***	***0.5***
aim ppm	*150*	*145.8*	*141.5*	*137.3*	*133.1*	*128.9*
NCP $	0.250	0.257	0.265	0.273	0.282	0.291
ECP	1.789	1.709	1.619	1.523	1.427	1.336
cost $	*0.447*	*0.439*	*0.429*	*0.416*	*0.402*	*0.389*

C_{pk}	***0.6***	***0.7***	***0.8***	***0.9***	***1.0***	***1.1***
aim ppm	*124.6*	*120.4*	*116.2*	*111.9*	*107.7*	*103.5*
NCP $	0.301	0.311	0.323	0.335	0.348	0.362
ECP	1.255	1.186	1.130	1.087	1.055	1.033
cost $	*0.378*	*0.369*	*0.365*	*0.364*	*0.367*	*0.374*

C_{pk}	***1.2***	***1.3***	***1.4***	***1.5***	***1.6***	***1.7***
aim ppm	*99.2*	*95.0*	*90.8*	*86.6*	*82.3*	*78.1*
NCP $	0.378	0.395	0.413	0.433	0.456	0.480
ECP	1.019	1.010	1.005	1.003	1.001	1.000
cost $	*0.385*	*0.399*	*0.415*	*0.434*	*0.456*	*0.480*

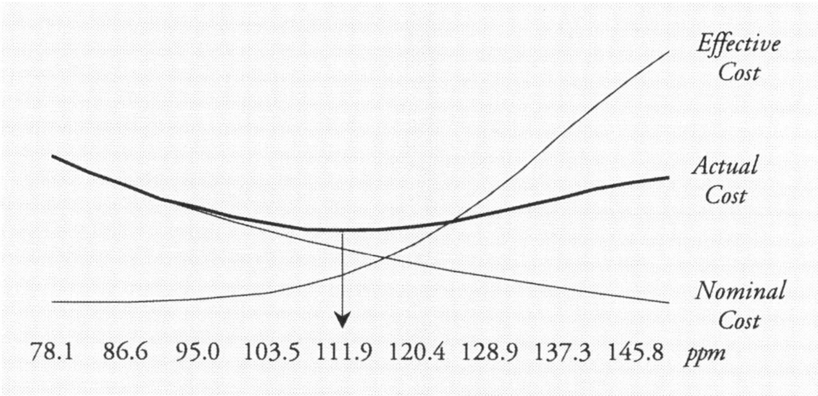

Figure 4.6: The Optimum Aim for Table 4.3

4.6 The Dividends of a Predictable Process

To assess the payback from operating your process predictably you will need to go through the same type of analysis that was used to determine the optimum aim point. However, in this analysis you will use the *predictable scale factor* obtained from a process behavior chart using the data collected earlier.

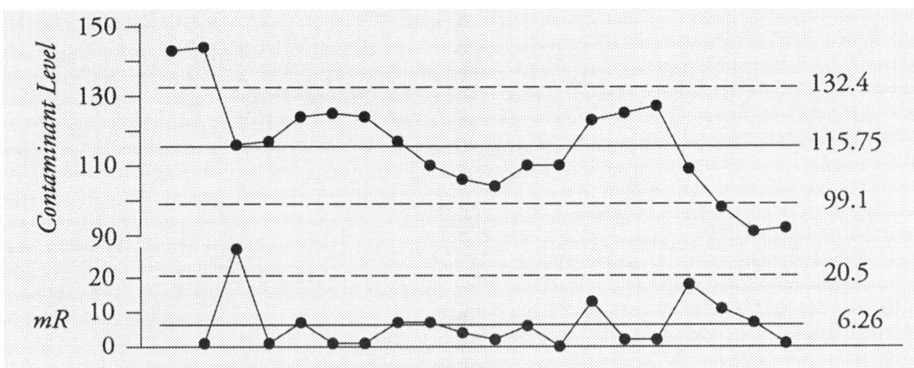

Figure 4.7: *XmR* Chart for Contaminant Level Data

Figure 4.7 shows the *XmR* Chart for the Contaminant Level Data. The data show that process is being operated in an unpredictable manner. Therefore we will want to know what can be gained if we learn how to operate it predictably.

The first step is to reassess the scale factor: the average moving range is 6.26, giving a predictable scale factor of:

$$\textit{predictable scale factor} = \frac{\bar{R}}{d_2} = \frac{6.26\ ppm}{1.128\ std.\ dev.} = 5.55\ ppm\ per\ std.\ dev.$$

To assess the benefits of operating the process predictably while leaving the process average unchanged we can compare the Effective Cost of Production with the Potential Cost of Production. The Effective Cost of Production found using the past performance data was 1.130. The Potential Cost of Production is found by using the predictable scale factor to compute a hypothetical centered capability ratio:

$$Hypothetical\ C_{pk} = \frac{150\ ppm - 115.75\ ppm}{3.0 \times 5.55\ ppm\ per\ s.d.} = 2.06$$

And this hypothetical C_{pk} value is off the page on Table 21. This means that the Potential Cost of Production is 1.000. Comparing this value with the value of 1.130 found earlier, we see that holding the process average steady at 116 ppm and operating it predictably should result in at least a thirteen percent drop in the cost of production, from 36.6 cents per pound to 32.4 cents, for an annual savings of $ 1.05 million.

But as we learn how to operate this process more consistently, we may well want to change the process average. Since operating a process predictably will reduce the scale factor, the evaluation of the optimum aim point shown in Table 4.3 is no longer appropriate. We will need to reevaluate the optimum aim point using the predictable scale factor.

With a new scale factor the same C_{pk} values will represent different possible aim points. For each of the C_{pk} values in Table 21 we use the predictable scale factor to compute new aim points:

$$aim\ point = specification \pm [\ C_{pk} \times scale\ factor \times 3.0\ ]$$

$$aim\ point\ in\ ppm = 150\ ppm - [\ C_{pk} \times 5.55\ ppm\ per\ std.\ dev \times 3.0\ ]$$

These aim points are shown in Table 4.4. Given these new aim points you can now (based on your process knowledge) compute the nominal cost for each aim point. In this case it was known that the energy costs for producing this compound, in dollars per pound, could be obtained from the formula:

$$cost\ per\ pound = \frac{37.5}{average\ level\ of\ contaminant\ in\ ppm}$$

Finally, the Effective Costs of Production (ECP) values are obtained from the second column of Table 21, multiplied by the nominal costs of production (NCP) to get the actual costs in dollars per pound (*cost*). These values are all shown in Table 4.4.

Table 4.4: Actual Costs for Contaminant Level Data

C_{pk}	*0.0*	*0.1*	*0.2*	*0.3*	*0.4*	*0.5*
aim ppm	*150*	*148.3*	*146.7*	*145.0*	*143.3*	*141.7*
NCP $	0.250	0.253	0.256	0.259	0.262	0.265
ECP	1.789	1.709	1.619	1.523	1.427	1.336
cost $	*0.447*	*0.432*	*0.414*	*0.394*	*0.374*	*0.354*
C_{pk}	*0.6*	*0.7*	*0.8*	*0.9*	*1.0*	*1.1*
aim ppm	*140.0*	*138.3*	*136.7*	*135.0*	*133.3*	*131.7*
NCP $	0.268	0.271	0.274	0.278	0.281	0.285
ECP	1.255	1.186	1.130	1.087	1.055	1.033
cost $	*0.336*	*0.321*	*0.310*	*0.302*	*0.296*	*0.294*
C_{pk}	*1.2*	*1.3*	*1.4*	*1.5*	*1.6*	*1.7*
aim ppm	*130.0*	*128.7*	*126.7*	*125.0*	*123.4*	*121.7*
NCP $	0.288	0.291	0.296	0.300	0.304	0.308
ECP	1.019	1.010	1.005	1.003	1.001	1.000
cost $	*0.293*	*0.294*	*0.297*	*0.301*	*0.304*	*0.308*

The minimum Actual Cost in Table 4.4 is $ 0.293 and this cost occurs in the vicinity of 130 ppm. Comparing the 29.3 cents per pound found here with the Past Performance Cost of 36.6 cents per pound, we see that the minimum payback we could expect from learning how to operate this process predictably at the optimum aim point is $ 0.073 per pound. At 25 million pounds annually, this amounts to an extra $ 1.825 million in profits for this operation each year.

4.7 Practice

A critical characteristic for a sheet of film is thickness. The 30 thickness values below were obtained from end-of-the-roll samples from the last roll produced each day. The measurements are given in mils.

5.3	4.5	5.1	4.5	5.2	4.5	4.9	5.4	4.7	4.6
4.0	2.3	1.6	1.9	1.5	2.3	1.6	1.9	1.5	5.8
5.7	5.7	5.2	4.6	4.8	3.5	3.5	3.4	4.6	5.0

The minimum thickness specification is 2.3 mils. Assume that the nominal cost of production varies in a linear manner and that the nominal cost at

2.3 mils is $ 1.53 per pound, while the nominal cost at 3.0 mils is $ 1.60 per pound. Nominal cost = $ [1.30 + (thickness in mils/10)]. Nonconforming product is downgraded and the resulting loss in revenue is equivalent to rework costing 50% of the nominal cost of production.

- Characterize the past performance of this process.
- Evaluate the potential payoff for improving the process aim.
- Evaluate the potential payoff from operating this process predictably and on target. (Place the 30 values on an *XmR* Chart.)

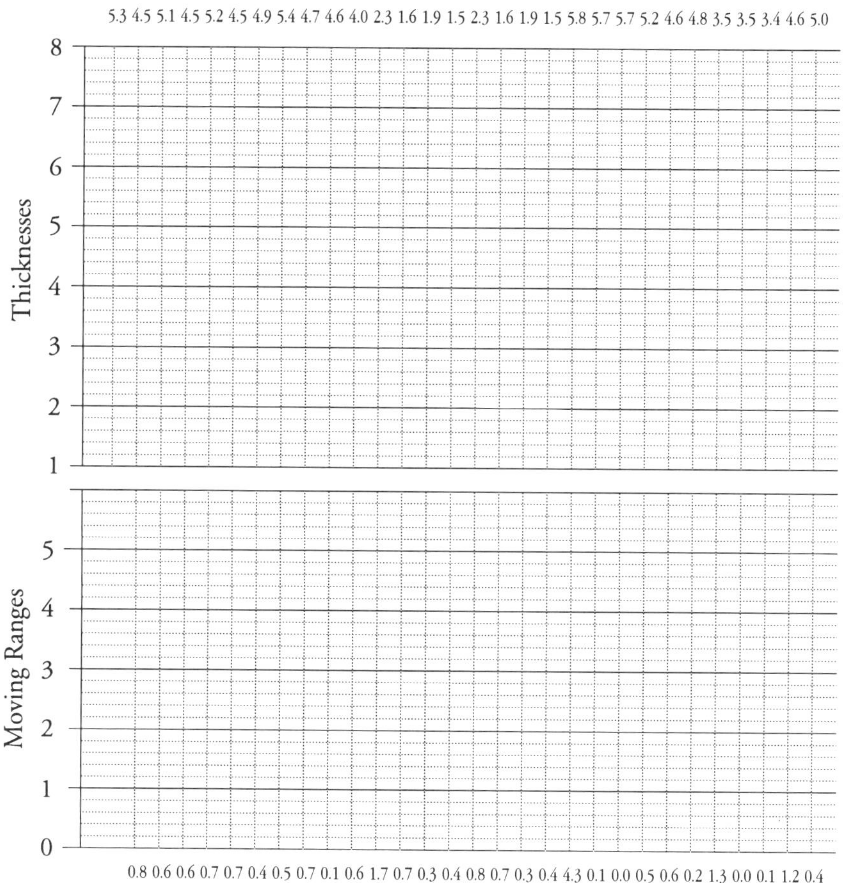

Worksheet for One-Sided Specification Without Target

Part One: Characterizing Past Performance

1. **Identify a critical product characteristic:**

   Product Identification: ______________________________

   Critical Characteristic Evaluated: ______________________________

   Specification Limit = SL = ____________

2. **Compute a *scale factor* using a global measure of dispersion:**

   Number of values used = N = ______

   scale factor: s = ____________ or R/d_2 = ____________

3. **Characterize the Distance to (Nearer) Specification:**

   average value for the N data = ____________

   $USL - average$ or $average - LSL$ = ____________

   $$DNS = \frac{average\ \ to\ spec}{scale\ factor} = \underline{\qquad\qquad} = \underline{\qquad\qquad}$$

   Divide the *DNS* value by 3.0 to obtain your

   Centered Performance Ratio, P_{pk} = [ ]

4. **Find the Effective Cost of Production:**

   Is nonconforming product scrapped or reworked?

   If it is reworked, then what is the cost of rework as a proportion of the nominal cost of production? 100% 80% 67% 50% 33% 20% 10%

   Use the appropriate column of Table 21, let P_{pk} determine the row, and read off your **Effective Cost of Production**: [ ]

   Multiply by the nominal cost of production (at the average) to find the **actual cost of production** for past production [ ]

Worksheet for One-Sided Specification Without Target

page 2 of 5

Part Two: Potential Payoff for Improving Process Aim

5. Find the aim points that correspond to the C_{pk} values of Table 21:

$$aim\ point = specification \pm [\ C_{pk} \times scale\ factor \times 3.0\ ]$$

C_{pk}	Aim Point	Nominal Cost	Effective Cost	Actual Cost
0.0	______	______	______	______
0.1	______	______	______	______
0.2	______	______	______	______
0.3	______	______	______	______
0.4	______	______	______	______
0.5	______	______	______	______
0.6	______	______	______	______
0.7	______	______	______	______
0.8	______	______	______	______
0.9	______	______	______	______
1.0	______	______	______	______
1.1	______	______	______	______
1.2	______	______	______	______
1.3	______	______	______	______
1.4	______	______	______	______
1.5	______	______	______	______
1.6	______	______	______	______
1.7	______	______	______	______

Worksheet for One-Sided Specification Without Target

page 3 of 5

6. **For each of the aim points find the nominal cost of production**
Use your process knowledge to obtain reasonable and appropriate estimates of the nominal cost of production at each one of the different aim points computed in Step 5 above.

7. **Use Table 21 to obtain the Effective Costs of Production**
Use the appropriate column of Table 21 to fill in the Effective Costs of Production that correspond to the C_{pk} values shown in Step 5.

8. **Multiply the nominal costs by the Effective Costs to find the actual costs of production**
Identify the aim point that gives the minimum actual cost.

Optimum aim point = ____________

Effective Cost of Production at optimum aim = ____________

Optimum actual cost of production = ____________

Compare this actual cost with the value found in Step 4 to determine the potential payback from improving the process aim.

Worksheet for One-Sided Specification Without Target

page 4 of 5

Part Three: Potential Payoff for Operating Predictably

9. Place your past performance data on a process behavior chart:

☐ If your process appears to be predictable, skip the remainder of this worksheet. Unless or until your process is changed in some major way, the cost of production in Step 4 is what you should expect from your process in the future. If you improve your process aim you can expect to get the cost of production found in Step 8.

☐ If your process appears to be unpredictable, then you will need to compute a *Potential Cost of Production*. As a first step toward this value use the Average Range, or the Average Moving Range, from your process behavior chart and the appropriate bias correction factor from Table 38 to compute a *predictable scale factor:*

subgroup size, n = _____ (use $n = 2$ for XmR chart)

Average Range = $\bar{R}$ = __________

$$\textit{predictable scale factor} = \frac{\bar{R}}{d_2} = \text{\_\_\_\_\_\_\_\_\_\_} = \text{\_\_\_\_\_\_\_\_\_\_}$$

10. Evaluate the benefits of operating predictably at current average:
Use the predictable scale factor above to compute a hypothetical centered capability ratio:

$$\textit{Hypothetical } C_{pk} = \frac{|\,\textit{Average} - \textit{Specification}\,|}{3 \times \textit{predictable scale factor}} = \text{\_\_\_\_\_\_\_\_} = \text{\_\_\_\_\_\_\_\_}$$

Use this value with the appropriate column of Table 21 to find a
Potential Cost of Production = __________
Compare with the Effective Cost of Production found in Step 4 to determine the benefits of operating this process predictably at the current average.

Worksheet for One-Sided Specification Without Target

page 5 of 5

11. Evaluate the benefits of operating predictably at optimum aim:

Recompute the aim points using the *predictable scale factor*:

$$aim\ point = specification \pm [\ C_{pk} \times scale\ factor \times 3.0\ ]$$

C_{pk}	Aim Point	Nominal Cost	Effective Cost	Actual Cost
0.0	______	______	______	______
0.1	______	______	______	______
0.2	______	______	______	______
0.3	______	______	______	______
0.4	______	______	______	______
0.5	______	______	______	______
0.6	______	______	______	______
0.7	______	______	______	______
0.8	______	______	______	______
0.9	______	______	______	______
1.0	______	______	______	______
1.1	______	______	______	______
1.2	______	______	______	______
1.3	______	______	______	______
1.4	______	______	______	______
1.5	______	______	______	______
1.6	______	______	______	______
1.7	______	______	______	______

- Find the nominal cost of production at each aim point.
- Find the Effective Costs of Production at each aim point.
- Multiply nominal costs by Effective Costs and find the minimum actual cost.
- Compare this minimum with the actual cost in Step 4 to determine the benefits of operating predictably at optimum aim.

Chapter Five

One-Sided Specifications With A Stated Target

The previous chapter considered one-sided product specifications that did not have a stated target value. There we used a generic excess cost function to compute the values in Table 21. Here we will look at those one-sided specifications that have an stated target value, and we will use this additional piece of information in computing values for the Effective Cost of Production.

5.1 Stated Targets

In Chapter Four we found that the excess cost function for a one-sided product specification would look like the one shown in Figure 5.1. On one side of the specification you will scrap or rework the product. On the other side there will be a zone of degraded functionality followed by a zone of full functionality.

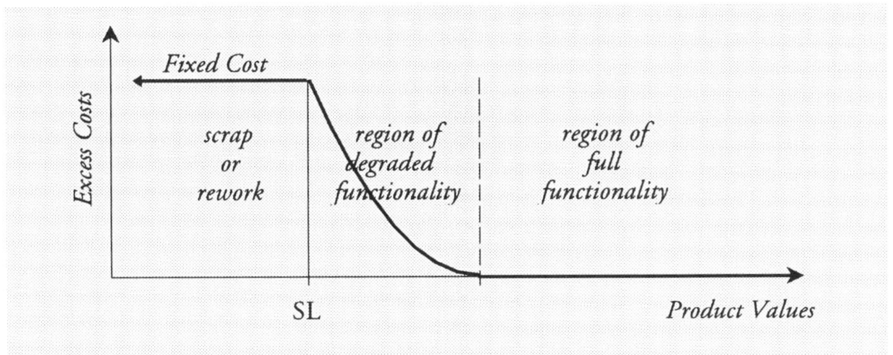

Figure 5.1: Excess Cost Function for a One-Sided Product Specification

The question in Chapter Four was where to draw the dashed line. There, by analogy with the two-sided specification case, we placed the line three standard deviations away from the specification. However, when a one-sided product specification has a stated target value it is reasonable to place the dashed line at this target value. The idea being that as you move from the target value toward the specification the functionality of the part will begin to degrade until, at the specification, the part becomes nonfunctional.

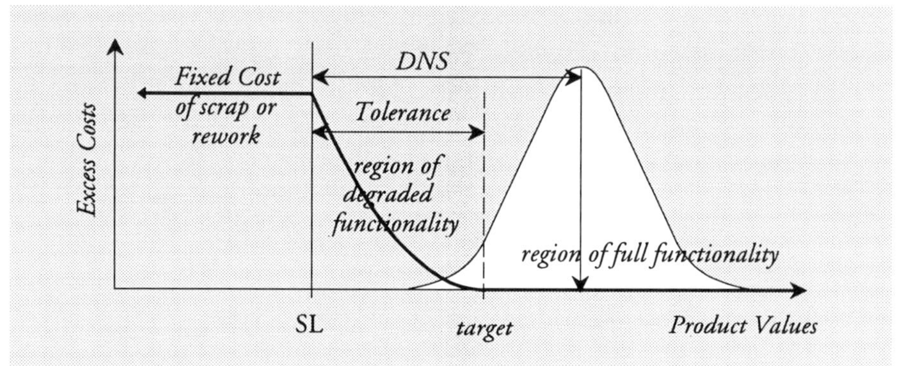

Figure 5.2: One-Sided Product Specification With Target Value

As you can see in Figure 5.2, the Effective Cost of Production will depend upon two distances. The first of these is the distance from the process average to the specification which is customarily referred to as the Distance to Nearer Specification (*DNS*). The second of these is the distance from the specification to the target value, which is customarily known as the *Tolerance*. (Obviously both of these terms come from the world of two-sided specifications. However, rather than introduce new terms for familiar concepts it is simpler to use these terms with one-sided specifications as well.) In Chapter Four this *Tolerance* was fixed at a value of three standard deviations in order to obtain a generic excess cost function. Here we will let the stated target value determine the *Tolerance* value.

The target value for any product characteristic should define that value that has zero excess costs—that point on the product value axis where, in a perfect world, you would like see every item to be produced. With a one-sided product specification the target is the minimum or maximum point that satisfies this condition. Of course, if it becomes difficult to define a target value you may always return to Chapter Four and use Table 21 for your process evaluation.

5.2 Characterizing Past Performance

The problem of identifying the target value for a one-sided product specification is not a statistical problem. It is an engineering problem, requiring process knowledge and judgment. Therefore it will be part of the assumed portion of the following examples.

Step One: Identify the critical characteristic, its specification and its target value. The way to begin any evaluation is to collect the known information about the process.

Step Two: Compute a scale factor using a global measure of dispersion. In doing this it is still best to use at least 50 data collected over an extended period of time. Twenty data are an absolute minimum.

Step Three: Express the Tolerance in standard deviation units. Obtain the distance between the specification and the target value and divide this distance by the scale factor to obtain the *Tolerance* statistic:

$$Tolerance = \frac{|\, specification - target \,|}{scale\ factor}$$

Tables 22 to 29 are set up to use this *Tolerance* value which is expressed in standard deviation units.

Step Four: Characterize the Distance to Nearer Specification. Obtain the distance between the average of your data and the specification. Divide this distance by the scale factor to obtain the Distance to Nearer Specification:

$$DNS = \frac{|\, average - specification \,|}{scale\ factor}$$

And then divide this value by 3.0 to obtain your centered performance ratio:

$$P_{pk} = \frac{DNS}{3.0}$$

Step Five: Determine the Effective Cost of Production. Choose the appropriate table from Tables 22 to 29. Use the *Tolerance* to choose the column, use the P_{pk} value to determine the row, and find the Effective Cost of Production.

The critical characteristic for many round parts is the out-of-round dimension. For Part 1411 the maximum allowable out-of-round was 5 mils. Analysis of this part had convinced both the supplier and the customer that parts with less than 3.0 mils out-of-round were fully functional. Assume that nonconforming parts are scrapped.

4.4	3.8	3.8	3.6	3.8	4.2	4.6	4.2	4.2	4.2
4.8	4.2	3.2	3.6	3.8	4.0	4.4	4.6	3.6	4.0
4.2	4.0	4.4	4.2	3.6	3.4	4.0	3.4	3.4	3.8
4.8	3.8	4.4	3.6	3.8	4.2	4.0	4.2	3.6	4.2

Global measures of dispersion are:

$$s = 0.392 \; \textit{mils per standard deviation unit,} \text{ and}$$

$$\frac{R}{d_2} = \frac{0.8 \; \textit{mils}}{4.32 \; \textit{std. dev.}} = 0.370 \; \textit{mils per standard deviation}$$

With a target of 3.0 mils and an upper specification limit of 5 mils the Tolerance is:

$$\textit{Tolerance} = \frac{3.0 \; \textit{mils} - 5.0 \; \textit{mils}}{0.392 \; \textit{mils per std dev}} = 5.10 \; \textit{standard deviations}$$

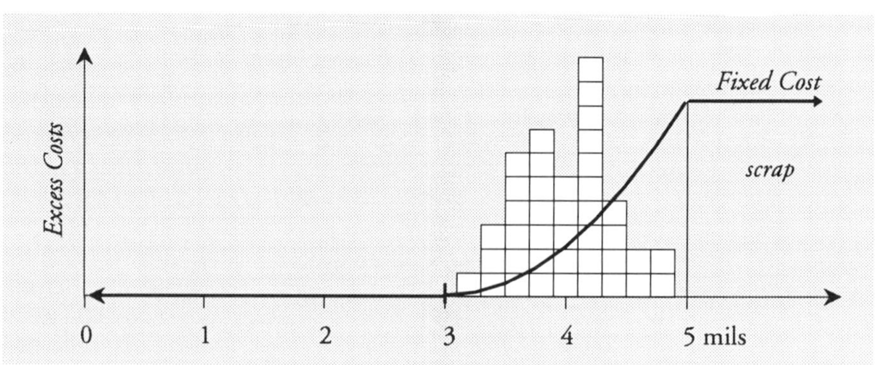

Figure 5.3: The Out-of-Round Data

These data have an average of 4.00. The Distance to Nearer Specification is:

$$DNS = \frac{4.00\ mils - 5.0\ mils}{0.392\ mils\ per\ std\ dev} = 2.55\ standard\ deviations$$

from which we find the centered performance ratio:

$$P_{pk} = \frac{5.1}{3.0} = 0.85$$

And using Table 22, with *Tolerance* = 5.1 and P_{pk} = 0.9 we get an Effective Cost of Production of 1.26. Thus, we would estimate that when P_{pk} is 0.9 the actual cost of production is about 26 percent more than the nominal cost of production.

At P_{pk} = 0.9, the nominal cost of production is about 35.5 cents per piece. Thus, we estimate the actual cost to be about 44.7 cents per piece.

5.3 Benefits of Improving the Process Aim

How would you determine the optimum aim point for Part 1411?

Step One: For the appropriate C_{pk} values from the table used in characterizing the past performance you should compute the corresponding aim points for the out-of-round dimension. The equation for this computation is:

$$aim\ point = specification \pm [\ C_{pk} \times scale\ factor \times 3.0\ ]$$

where the plus is used with minimum specifications and the minus is used with maximum specifications. Thus, for this example:

$$Out\text{-}of\text{-}Round\ in\ mils = 5\ mils - [\ C_{pk} \times 0.392\ mils\ per\ std.\ dev \times 3.0\ ]$$

Using C_{pk} values that range from 0.0 to 2.0 we get the aim points shown in Table 5.1.

Step Two: Determine the nominal cost at the different aim points. The nominal cost of production is related to the out-of-round dimension by the expression:

$$cost\ in\ cents\ per\ piece = \frac{140}{average\ out\text{-}of\text{-}round\ in\ mils}$$

Step Three: Obtain the Effective Costs of Production from the appropriate table using the *Tolerance* value computed earlier. Multiply by the nominal costs to obtain the actual costs at the different aim points.

Table 5.1: Finding the Optimum Aim for Part 1411

C_{pk}	0.0	0.1	0.2	0.3	0.4	0.5	0.6
aim mils	*5.0*	*4.88*	*4.77*	*4.65*	*4.53*	*4.41*	*4.29*
NCP	28	28.7	29.4	30.1	30.9	31.7	32.6
ECP	2.73	2.31	2.03	1.83	1.68	1.56	1.47
cost	*76.4*	*66.2*	*59.6*	*55.1*	*51.9*	*49.5*	*47.9*

C_{pk}	0.7	0.8	0.9	1.0	1.1	1.2	1.3
aim mils	*4.18*	*4.06*	*3.94*	*3.82*	*3.71*	*3.59*	*3.47*
NCP	33.5	34.5	35.5	36.6	37.8	39.0	40.3
ECP	1.39	1.32	1.26	1.21	1.16	1.12	1.09
cost	*46.6*	*45.5*	*44.8*	*44.3*	*43.8*	*43.7*	*44.0*

C_{pk}	1.4	1.5	1.6	1.7	1.8	1.9	2.0
aim mils	*3.35*	*3.24*	*3.12*	*3.00*	*2.88*	*2.77*	*2.65*
NCP	41.7	43.3	44.9	46.7	48.6	50.6	52.9
ECP	1.07	1.05	1.03	1.02	1.01	1.01	1.00
cost	*44.7*	*45.4*	*46.2*	*47.6*	*49.0*	*51.1*	*52.9*

Step Four: Find the aim point that corresponds to the minimum actual cost, and compare the cost at that point with the actual cost from the characterization of past performance. For Part 1411 the minimum cost of 43.7 cents per piece occurs at an aim point of 3.6 mils. Compared with the 44.7 cents found earlier it would appear that they could save about one cent per piece. While they have been operating near C_{pk} = 0.9 the optimum aim appears to be about C_{pk} = 1.2. At this optimum aim point the nominal cost would appear to be 3.5 cents per piece greater, but the excess costs associated with the degraded functionality of parts near the specification would be about 4.6 cents less.

So what can be accomplished by improving the process aim for Part 1411? About 1 cent per part is all that can be gained by improved process aim. At 2500 parts per shift this amounts to about $ 5000 in a year. With such a small potential payback only simple things need to be tried here.

5.4 Benefits of Predictable Operation

To assess the potential payoff from operating a process predictably and at the optimum aim point we will need to characterize the past behavior of the process. And the only way to do this is to place your data on a process behavior chart. Figure 5.4 shows an *XmR* chart for the out-of-round dimensions for Part 1411.

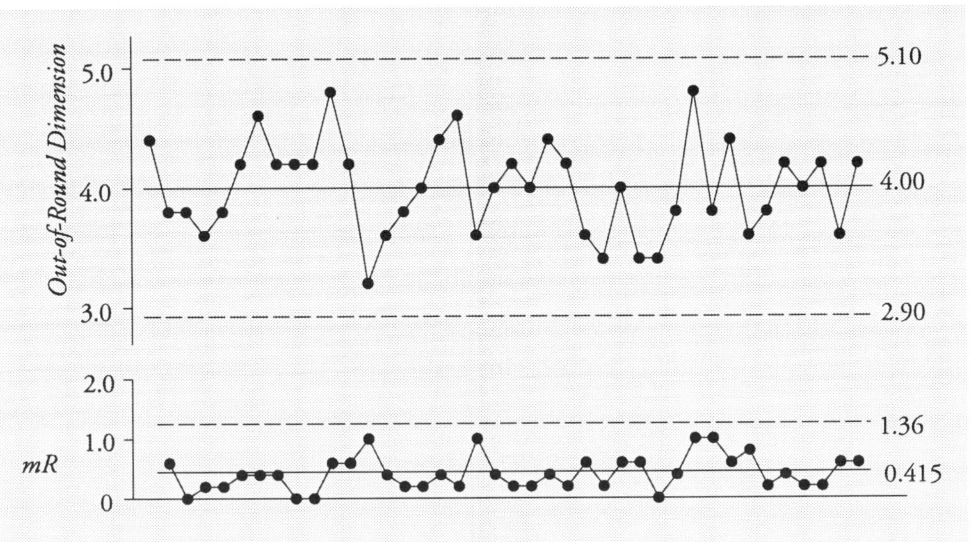

Figure 5.4: *XmR* Chart for Part 1411

There is no evidence of unpredictable behavior in these past performance data. This process appears to be reasonably predictable, and as we found in the previous section, it is already in the neighborhood of the optimum aim point. Since a predictable process does not occur by chance, it is likely that they already have an effective operational strategy in place for this process. You are going to have to find another process to improve in order to look like a hero!

If the process appears to be unpredictable, then you would go on to Step Two: Compute a predictable scale factor. If you do this with the data for Part 1411 you will find essentially the same value we found when characterizing the past performance.

Step Three would be to compute a hypothetical C_{pk} value based on the average and the predictable scale factor. Then, using the *Tolerance* value found earlier, this hypothetical C_{pk} could be use to obtain a Potential Cost of Production. By comparing this Potential Cost of Production with the Effective Cost of Production based on past performance you can assess the potential payoff for operating predictably at the current average value.

To assess the potential payoff for operating predictably at the optimum aim you will need to recompute the optimum aim worksheet using the predictable scale factor to find new aim points. Comparison of the potential minimum actual cost value from this table would then be compared with the actual cost estimated in the characterization of past performance. The difference would be a lower bound on the potential savings from operating your process predictably at the optimum aim.

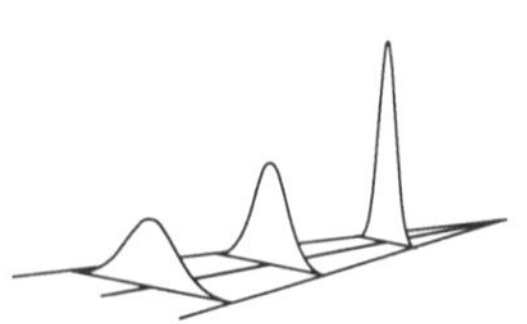

Worksheet for a One-Sided Specification With Target

Part One: Characterizing Past Performance

1. Identify a critical product characteristic:

product identification: ____________________

critical characteristic evaluated: ____________________

specification limit = *SL* = ____________

target value = *target* = ____________

2. Compute a *scale factor* using a global measure of dispersion:

number of values used = *N* = ______

scale factor: s = ____________ or R/d_2 = ____________

3. Characterize the *Tolerance* in standard deviation units:

$$Tolerance = \frac{|\ target - SL\ |}{scale\ factor} = \frac{\qquad}{\qquad} = \boxed{\qquad}$$

4. Characterize the Distance to (Nearer) Specification:

average value for the *N* data = ____________

USL – average or *average – LSL* = ____________

$$DNS = \frac{average\ \ to\ spec}{scale\ factor} = \frac{\qquad}{\qquad} = \underline{\qquad}$$

Divide the *DNS* value by 3.0 to obtain your

Centered Performance Ratio, P_{pk} = $\boxed{\qquad}$

Worksheet for a One-Sided Specification With Target

page 2 of 6

5. **Find the Effective Cost of Production:**

Is nonconforming product scrapped or reworked?

If it is reworked, then what is the cost of rework as a proportion of the nominal cost of production? 100% 80% 67% 50% 33% 20% 10%

Find the appropriate table from Tables 22 through 29: Table _____

Use the *Tolerance* value and the P_{pk} value to find your **Effective Cost of Production**: [ ]

Multiply by the Nominal Cost of Production (at the average) to find the **actual cost of production** for past production [ ]

Part Two: Potential Payoff for Improving Process Aim

6. **Find aim points that correspond to the C_{pk} values of the table above:**

Fill these values in on the following page.

$$aim\ point = specification \pm [\ C_{pk} \times Scale\ Factor \times 3.0\ ]$$

7. **For each of the aim points find the nominal cost of production**
Use your process knowledge to obtain reasonable and appropriate estimates of the nominal cost of production at each of the different aim points computed in Step 6 above.

8. **Use the table above to obtain the Effective Costs of Production**
Use the appropriate column of the table identified in Step 5 to fill in the Effective Costs of Production that correspond to the C_{pk} values shown on the next page.

Worksheet for a One-Sided Specification With Target

page 3 of 6

C_{pk}	Aim Point	Nominal Cost	Effective Cost	Actual Cost
0.0	______	______	______	______
0.1	______	______	______	______
0.2	______	______	______	______
0.3	______	______	______	______
0.4	______	______	______	______
0.5	______	______	______	______
0.6	______	______	______	______
0.7	______	______	______	______
0.8	______	______	______	______
0.9	______	______	______	______
1.0	______	______	______	______
1.1	______	______	______	______
1.2	______	______	______	______
1.3	______	______	______	______
1.4	______	______	______	______
1.5	______	______	______	______
1.6	______	______	______	______
1.7	______	______	______	______
1.8	______	______	______	______
1.9	______	______	______	______
2.0	______	______	______	______

Worksheet for a One-Sided Specification With Target

9. On page 3 multiply the nominal costs by the Effective Costs to find the actual costs of production

Identify the aim point that gives the minimum actual cost.

Optimum aim point = ____________

Optimum actual cost of production = ____________

To determine the potential payback from improving the process aim compare this actual cost with the value found in Step 5.

Part Three: Potential Payoff for Operating Predictably

10. Place your past performance data on a process behavior chart:

☐ If your process appears to be predictable, skip the remainder of this worksheet. Unless or until your process is changed in some major way, the cost of production in Step 5 is what you should expect from your process in the future. If you improve your process aim you can expect to get the cost of production found in Step 9.

☐ If your process appears to be unpredictable, then you will need to compute a *Potential Cost of Production*. As a first step toward this value use the Average Range, or the Average Moving Range, from your process behavior chart and the appropriate bias correction factor from Table 38 to compute a *predictable scale factor:*

subgroup size, n = ______ (use $n = 2$ for XmR chart)

Average Range = $\bar{R}$ = ____________

predictable scale factor = $\frac{\bar{R}}{d_2}$ = ____________ = ____________

Worksheet for a One-Sided Specification With Target

page 5 of 6

11. Evaluate the benefits of operating predictably at current average:
Use the predictable scale factor above to compute a hypothetical centered capability ratio:

$$Hypothetical\ C_{pk} = \frac{|\ Average - Specification\ |}{3 \times predictable\ scale\ factor} = \_\_\_\_\_\_\_\_\_\_$$

Use this value along with the *Tolerance* value found in Step 3 in the table identified in Step 5 to find a

Potential Cost of Production = __________

Compare with the value in Step 5 to determine the benefits of operating this process predictably at the *current* average.

Part Four: Payoff for Predictable at Optimum Aim

12. Recompute the aim points:
Using the *predictable scale factor* from Step 10 find the aim points that correspond to the C_{pk} values on the next page:

$$aim\ point = specification \pm [\ C_{pk} \times scale\ factor \times 3.0\ ]$$

13. Find the nominal costs at these aim points:

14. Find the Effective Costs of Production: Use the table identified in Step 5 and the *Tolerance* identified in Step 3 with the C_{pk} values shown. Multiply the Effective Costs by the nominal costs.

15. Find the potential payoff for operating predictably at optimum aim:
Compare the minimum actual cost with the values found in Step 5 to estimate the minimum payoff for operating predictably at the optimum aim point.

Worksheet for a One-Sided Specification With Target

page 6 of 6

C_{pk}	Aim Point	Nominal Cost	Effective Cost	Actual Cost
0.0	______	______	______	______
0.1	______	______	______	______
0.2	______	______	______	______
0.3	______	______	______	______
0.4	______	______	______	______
0.5	______	______	______	______
0.6	______	______	______	______
0.7	______	______	______	______
0.8	______	______	______	______
0.9	______	______	______	______
1.0	______	______	______	______
1.1	______	______	______	______
1.2	______	______	______	______
1.3	______	______	______	______
1.4	______	______	______	______
1.5	______	______	______	______
1.6	______	______	______	______
1.7	______	______	______	______
1.8	______	______	______	______
1.9	______	______	______	______
2.0	______	______	______	______

Chapter Six

Targets at Boundaries

In Chapters One, Two, and Three we considered the case of two-sided specifications. In Chapters Four and Five we looked at one-sided specifications. In this chapter we will consider that hybrid situation where the product characteristic is bounded on one side, has a specification limit on the other side, and has a target that corresponds to the boundary condition.

6.1 The Excess Cost Function

When the target value coincides with a boundary condition we will have a one-sided excess cost function as shown in Figure 6.1. As before, we will assume that the excess costs are zero at the target value. We will also assume that there will be a fixed cost of scrap or rework for nonconforming items. In between the zero cost and the fixed cost we will once again use a quadratic curve to characterize the excess costs associated with conforming product that deviates from the target value.

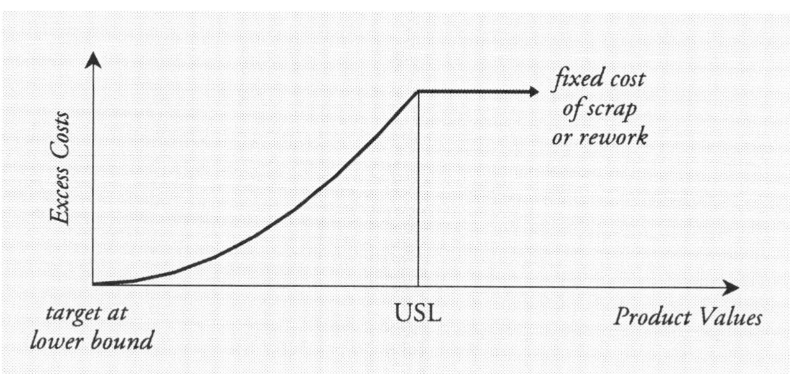

Figure 6.1: Excess Cost Function when the Target is at a Boundary

Of course a boundary condition places a restriction on the possible values for the product characteristic. To incorporate this restriction into our computations we will need to use a bounded probability model such as the one shown in Figure 6.2 to compute the excess costs.

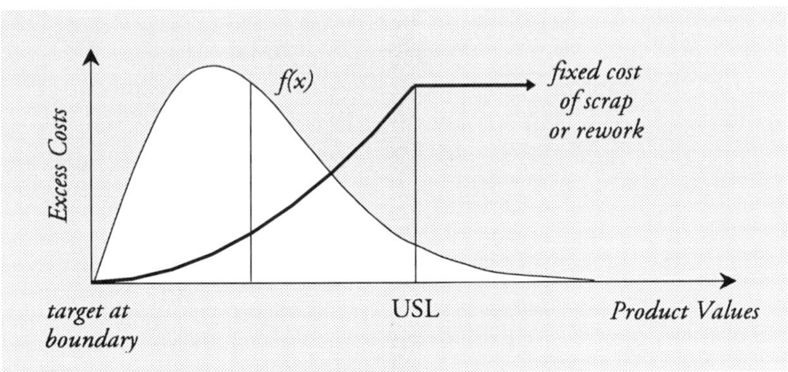

Figure 6.2: A Probability Model for when the Target is at a Boundary

If we let C denote the fixed cost of scrap or rework, the excess cost function for Figure 6.2 would look like:

$$\textit{Excess Costs} \;=\; C \int_{target}^{USL} \frac{(x - target)^2}{(USL - target)^2} f(x)\,dx \qquad \text{due to deviations above target}$$

$$+\; C \int_{USL}^{\infty} f(x)\,dx \qquad \text{excess costs due to rework or scrap}$$

Using the notation from Chapter One the expression above becomes:

$$\textit{Excess Costs} = C\,[\,\mathbf{IAT} + \mathbf{IRP}\,]$$

In order to incorporate this expression into the actual cost of production we will need to factor out the nominal cost of production, *NCP*, giving:

$$\textit{Excess Costs} = NCP \times \left\{ \frac{C}{NCP}\,[\,\mathbf{IAT} + \mathbf{IRP}\,] \right\}$$

And, as in Chapter One, the Effective Cost of Production will have the form:

$$\text{Effective Cost of Production} = \frac{1.0 - \text{proportion scrapped} + \text{Curly Brackets}}{1.0 - \text{proportion scrapped}}$$

While this equation is the same as the one given in Chapter One, the structure of the *Curly Brackets* expression has changed slightly. It will depend upon the integral above the target, the integral for the nonconforming proportion, and the ratio between the fixed cost of nonconforming product and the nominal cost of production.

In creating the tables of the Effective Cost of Production we could use the traditional capability indexes. However, because of the special features of the Target at Boundary case we can end up with C_{pk} values that are up to twice as large as the corresponding C_p value. Since this can create confusion, I have chosen to index Tables 30 to 37 using the Elbow Room and Distance to Nearer Specification values instead of the capability ratios. The Elbow Room will be the distance between the target (or boundary) and the specification limit, expressed in standard deviation units:

$$\text{Elbow Room} = \frac{|\ \text{specification} - \text{target}\ |}{\text{scale factor}}$$

And, depending upon the type of scale factor used, this value will characterize either the past performance or the potential process capability.

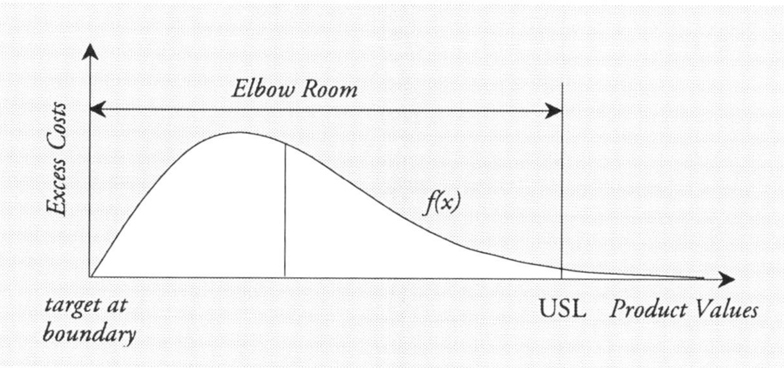

Figure 6.3: Elbow Room for Target at Boundary Case

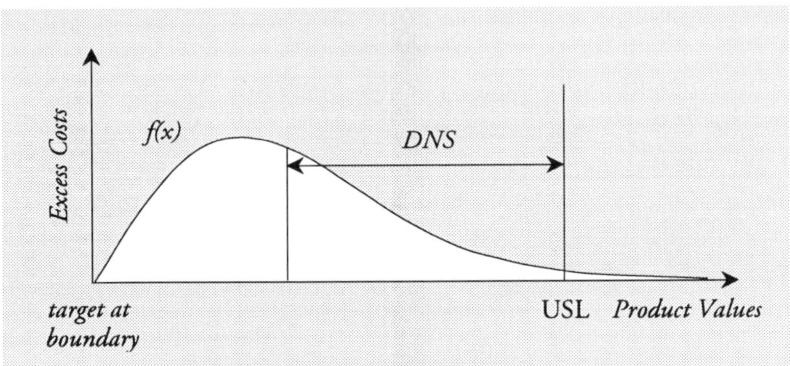

Figure 6.4: *DNS* for Target at Boundary Case

The Distance to Nearer Specification will be the distance from the process average to the specification value. This will always be the case even though the boundary may be closer to the average than the specification. (While boundary values are sometimes listed as specifications, they are not the same as a specification limit, and therefore are not appropriate for computing the *DNS* value.)

$$DNS = \frac{|\ specification - average\ |}{scale\ factor}$$

As always, when the average is on the wrong side of the specification limit the *DNS* value is assigned a minus sign to signify this condition.

6.2 Computing the Effective Costs of Production

In computing the Effective Costs of Production we can work with a lower bound. The values will apply equally well to those cases with an upper bound. Also, without loss of generality, we can let the lower bound be zero. The curve shown in Figure 6.4 is from the family of Burr distributions. This family of distributions has a probability distribution function of the form:

$$f(x) = c\,k\,x^{c-1}\,[\,1 + x^c\,]^{-k-1} \qquad \text{for} \quad x \geq 0, \;\; c > 0, \;\; k > 0$$

These distributions have a fixed left hand end-point, yet for different values of c and k they cover a wide region of the shape characterization plane. Thus they provide satisfactory mathematical models for use in evaluating the Curly Brackets expression. The mean and variance parameters for a Burr distribution may be found using the following formulas:

$$\mu = \frac{\Gamma(\frac{1}{c})\,\Gamma(k-\frac{1}{c})}{c\,\Gamma(k)} \qquad\qquad \sigma^2 = \frac{2\,\Gamma(\frac{2}{c})\,\Gamma(k-\frac{2}{c})}{c\,\Gamma(k)} - \mu^2$$

All distributions that have a fixed end-point will have some relationship between the mean and standard deviation. This means that for any one Burr distribution there will be a fixed relationship between the Elbow Room and the *DNS* values. However, by choosing different pairs of values for c and k we can get different combinations of Elbow Room and *DNS* values. This will allow us to create a set of tables for the Effective Cost of Production for situations having the target at a boundary value.

For a given Burr distribution, when *DNS* = 0 the mean will be equal to the specification limit, and the Elbow Room will be equal to the ratio of mean to the standard deviation for the Burr distribution. For other values of *DNS* the Elbow Room will be equal to *DNS* plus the ratio of mean to standard deviation.

$$\text{Elbow Room} = DNS + \frac{\text{mean of distribution}}{\text{standard deviation of distribution}}$$

Thus a single Burr distribution can be used to compute the Effective Costs of Production for a diagonal through the tables. By using appropriate Burr distributions the tables were built up, one diagonal at a time. The 40 Burr distributions used to compute the Effective Costs of Production shown in Tables 30 to 37 are listed in Table 6.1, along with the ratio of the mean to the standard deviation, the skewness, and the kurtosis of each distribution. Inspection of the values in Table 6.1 and consideration of Figure 6.4 will suggest that the values computed in Tables 30 to 37 should eventually converge to those in Tables 1, 2, 6, 8, 10, 12, 14, and 16.

Table 6.1: Burr Distributions Used for Tables 30 to 37

c	*k*	*mean* / *std. dev.*	*skewness*	*kurtosis*
0.41	35.65	0.3	14.2	545.7
0.64	67.50	0.6	4.38	37.5
0.93	34.75	0.9	2.45	12.8
1.22	72.83	1.2	1.54	6.41
1.55	58.55	1.5	1.07	4.43
1.89	61.26	1.8	0.75	3.53
2.24	64.45	2.1	0.52	3.09
2.60	60.25	2.4	0.35	2.88
2.97	48.90	2.7	0.21	2.78
3.33	64.07	3.0	0.09	2.74
3.70	66.00	3.3	0.00	2.74
4.08	53.50	3.6	-0.08	2.77
4.45	62.40	3.9	-0.15	2.81
4.83	57.00	4.2	-0.21	2.86
5.20	71.50	4.5	-0.26	2.91
5.58	68.76	4.8	-0.31	2.97
5.96	68.20	5.1	-0.35	3.02
6.34	68.80	5.4	-0.39	3.08
6.74	46.62	5.7	-0.42	3.14
7.10	73.30	6.0	-0.45	3.19
7.50	50.66	6.3	-0.48	3.24
7.88	53.25	6.6	-0.50	3.30
8.26	56.25	6.9	-0.53	3.35
8.64	59.66	7.2	-0.55	3.39
9.04	46.45	7.5	-0.57	3.44
9.42	49.22	7.8	-0.59	3.48
9.80	52.30	8.1	-0.61	3.52
10.18	55.75	8.4	-0.62	3.57
10.55	70.30	8.7	-0.64	3.61
10.94	63.80	9.0	-0.66	3.65
11.34	51.55	9.3	-0.67	3.68
11.72	54.90	9.6	-0.68	3.72
12.10	58.66	9.9	-0.69	3.75
12.48	62.80	10.2	-0.71	3.78
12.85	78.90	10.5	-0.72	3.82
13.25	62.90	10.8	-0.73	3.85
13.65	52.75	11.1	-0.74	3.87
14.05	45.77	11.4	-0.74	3.90
14.45	40.67	11.7	-0.75	3.92
14.80	56.80	12.0	-0.76	3.95

To see this convergence between the values in Tables 30 to 37 and Tables 1 to 16, you will have to divide both the Elbow Room and the *DNS* values by 3.0 to find the appropriate entries in Tables 1 to 20. Those columns in Tables 30 to 37 that have Elbow Room values that are greater than 12 were transcribed from the appropriate portions of Tables 1, 2, 6, 8, 10, 12, 14, and 16. This was done to provide interpolation points for large values of Elbow Room and *DNS*.

Tables 30 to 37 do not include negative *DNS* values. Approximate values of the Effective Cost of Production for negative DNS values may be obtained from Tables 1, 2, 6, 8, 10, 12, 14, and 16. Instructions for doing this are included in Tables 30 to 37.

In using Tables 30 to 37 the careful observer will note some small discrepancies between the values computed from the normal distribution and those obtained from the Burr distributions. This is to be expected—different probability models will result in different values. However, in most cases these differences will not substantially change the results. The conclusions as to which course of action will offer the greatest payback will remain the same in spite of these differences.

6.3 Characterizing Past Performance

Having identified a critical product characteristic, the specification limit, and the target value, you can begin by characterizing the past performance of your process. You will need a reasonable amount of data for this critical characteristic, preferably 50 or more values and at least 20 values in order to compute a global measure of dispersion. Using this global measure of dispersion as your scale factor, compute the Elbow Room and *DNS* values according to the formulas shown in Section 6.1.

Table 30 is for the case where the nonconforming product is scrapped. Table 31 is for the case where the nonconforming product is reworked at 100% of the nominal cost. Tables 32 to 37 are for rework that costs varying amounts relative to the nominal cost. Using the appropriate table, and the Elbow Room and DNS values found earlier, look up the Effective Cost of Production for past performance.

As an example of this procedure we will use the percentage, by weight, of Compound B in Product 927. Product 927 is made in a continuous process that produces about 500,000 pounds per annum. Compound B is a by-product of the reaction that is part of the process for producing Product 927. The target value for Compound B is 0%, and the upper specification limit for Compound B is 7% or 70 *parts per thousand*. Periodically a sample is taken and the amount of Compound B by-product is determined. When a nonconforming test value is found all of the product produced since the previous test is scrapped. Table 6.2 gives 95 of these measurements in parts per thousand arranged in time order by rows.

Table 6.2: 95 Values of Compound B in Product 927

21	28	40	37	44	44	58	63	56	50	47	19	38	72	59
62	25	28	35	42	43	61	27	45	46	38	36	41	49	52
58	53	51	44	35	40	41	40	44	42	52	46	48	37	39
38	60	42	41	42	45	33	35	49	29	25	37	49	45	42
36	27	32	30	34	30	28	75	25	31	54	52	39	48	29
32	30	31	39	35	39	75	54	52	38	35	54	52	37	60
56	50	50	42	49										

The average of the 95 values in Table 6.2 is 42.8 *ppt*. The standard deviation statistic is s = 11.54 *ppt per standard deviation unit*. Using these values to characterize past performance the Elbow Room is:

$$Elbow\ Room = \frac{70\ ppt}{11.54\ ppt/s.d.} = 6.07\ standard\ deviations$$

while the distance from the specification limit to the average is:

$$DNS = \frac{70\ ppt - 42.8\ ppt}{11.54\ ppt/s.d.} = 2.36\ standard\ deviations$$

From Table 30, rounding the Elbow Room to 6.0 and the *DNS* to 2.4, we find the Effective Cost of Production to be approximately 139 percent of the nominal cost.

6.4 Assessing Process Aim Strategies

When the target is at a boundary it will be impossible to operate exactly on target. Variation in the process outcomes will guarantee that the process average will always fall some distance away from the boundary. For this reason the adjustment of the process aim will seldom be as simple as it is in those cases where the process values do not impinge upon a barrier or boundary condition.

However, the excess cost function of Figure 6.1 makes it clear that the Effective Cost of Production will drop as the process average approaches the boundary target value. So we will want to look at ways of moving the process aim closer to the target value.

One obstacle to moving the process aim toward the target is the possibility of increasing costs near the target. If the nominal cost of production increases as the average moves toward the target then there will be some balance point between the competing forces of the excess cost function and the nominal cost function. As was shown in Chapter Four, you can determine this balance point by using the actual costs of production at various points. This balance point will be the optimum aim point for such a process.

In other cases the lack of a known mechanism for adjusting the process aim will result in a situation where the process aim can be moved only by making major process changes.

For these reasons, assessing the benefits of changes in the process aim before you have some idea of *how* to change the process aim may be a bit premature.

Since the nominal cost of production for Product 927 is the same regardless of the average amount of Compound B by-product, the objective remains to get the percentage of Compound B as close to zero as possible. Since we cannot simply adjust the amount of by-product produced, we look at the benefits of predictable operation.

6.5 The Dividends of a Predictable Process

The process behavior chart for the amount of Compound B in Product 927 is shown in Figure 6.5. The limits are based on the median moving range of 6.0 *ppt*. This chart shows that the process is being operated unpredictably. Therefore it will be interesting to estimate the Predictable Cost of Production and compare it with the Effective Cost of Production found earlier. With a median moving range of 6.0 our *predictable scale factor* is:

$$\frac{\textit{median moving range}}{d_4} = \frac{6}{0.954} = 6.3 \textit{ parts per thousand}$$

which results in predictable Elbow Room of 11.1 standard deviations, and a predictable *DNS* value of 4.32 standard deviations. Rounding this latter value off to 4.2, and using Table 30, we find the Predictable Cost of Production to be approximately 140 percent of nominal—which is essentially the same as was found earlier. *Without a change in process location there will be no substantial improvement in the Effective Cost of Production.* Work on operating this process predictably *at the current average* will not result in major gains.

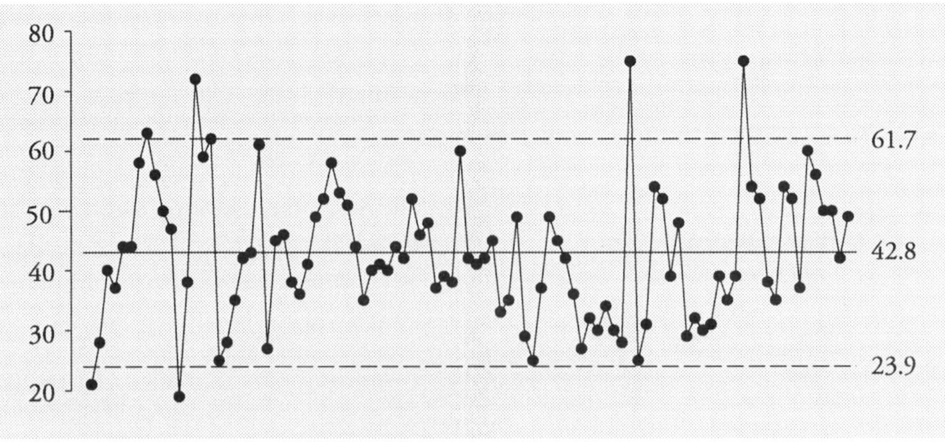

Figure 6.5: Amounts of Compound B in Product 927 for Samples 1 to 95

Efforts at improving this process need to be focused on how to reduce the average amount of Compound B.

6.6 The Reengineering Dividend

There is no simple mechanism for changing the average amount of Compound B in Product 927, and learning how to operate this process predictably at the current average offers no real advantage at the present time, so it is time to look at reengineering the process. The existing mechanisms for controlling the amount of Compound B were not very effective, so new mechanisms were sought. As a result of this project static mixers were installed in the feed streams. The effect of this change can be seen in the next 30 values.

Table 6.3: 30 Measurements of Compound B in Product 927 After Change

21	25	30	26	24	25	16	21	23	24	28	21	20	20	25
23	25	28	22	23	17	15	19	32	30	31	25	25	21	20

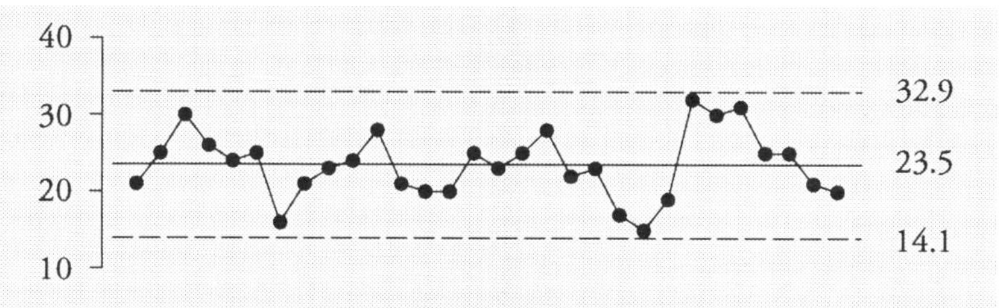

Figure 6.6: Amounts of Compound B in Product 927 for Samples 96 to 125

The average amount of Compound B in the next 30 samples was 23.5 *ppt*. The average moving range was 3.55 *ppt*. While these 30 values are much better behaved than the 95 values shown earlier, there is still some evidence of unpredictability in this process. In particular, samples 21 and 22 constitute a low run, while samples 24 to 26 constitute a high run. However, based on the change in the average amount of Compound B, the static mixers seem to have had a beneficial effect upon this process.

Assessing the Effective Cost of Production after this process change we would have a scale factor of:

$$\frac{\textit{average moving range}}{d_2} = \frac{3.55}{1.128} = 3.15 \textit{ units per std. dev.}$$

which gives an Elbow Room of 22.2 and a *DNS* of 14.76. Table 30 shows these values correspond to an Effective Cost of Production of approximately 114%. Thus, by reducing the amount of Compound B in Product 927 the process modification has netted a savings of about 25% of the nominal cost of production. However, the unpredictability of this process means that we should not expect these past savings to continue.

6.7 The Role of Predictable Operation

Figure 6.7 shows the amounts of Compound B in the next 30 samples of Product 927. The limits are those from Figure 6.6. The hints of unpredictability in Figure 6.6 have now matured into a highly erratic process.

Table 6.4: 30 Additional Measurements of Compound B in Product 927

24	26	48	29	20	29	32	39	40	55	38	40	54	56	35
28	29	23	44	23	26	31	26	39	29	34	36	30	39	45

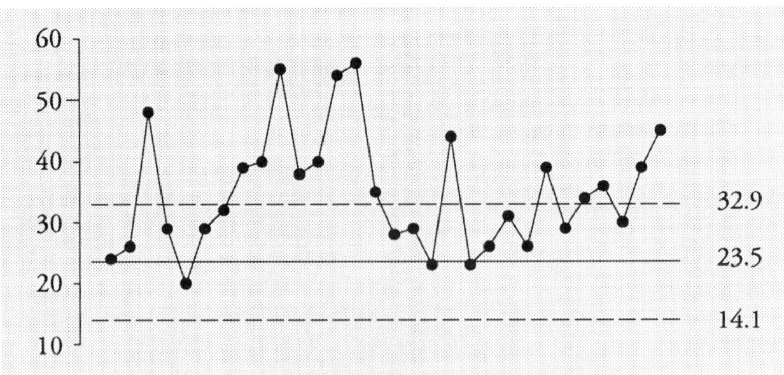

Figure 6.7: Amounts of Compound B in Product 927 for Samples 126 to 155

For the period covered by Figure 6.7 this process averaged 34.9 *ppt* of Compound B, with a standard deviation statistic of $s = 9.8$ *ppt*. Using these values to characterize Samples 126 to 155, we find an Effective Cost of Production of 127%. While this is better than when they started, it is not as good as this process is capable of doing.

When a process is not operated predictably it will be impossible to maintain any gains you might make. An unpredictable process is one that can change without warning. These changes are due major cause-and-effect relationships which influence the process outcomes and which are not being controlled. These relationships are known as assignable causes, and they will continue to dominate a process until they are identified and steps are taken to remove their influence over the process.

So, while they were able to improve things a bit with the static mixers in the feed streams, the benefits of this change were lost as other unknown assignable causes made their influence known.

Figure 6.6 shows that this process is capable of operating with an Effective Cost of Production of 114%. Figure 6.7 shows that this process is still subject to assignable causes of exceptional variation.

If they do not learn how to operate this process predictably, then they cannot count on even maintaining an Effective Cost of Production of 127%. (In the next section you will get to see what actually happened.)

A predictable process is one where the past can be used to characterize the future. An unpredictable process is one where the past will not characterize the future, except in the sense that the future will probably be worse than the past. In order to maintain gains of various process improvements you will need to operate your process predictably.

6.8 Practice

The next 30 samples of Product 927 had the following amounts of Compound B. The amounts are in parts per thousand. The values are written in time order by rows. The target is zero, and the specification limit is 70 *ppt*.

Table 6.5: Amounts of Compound B in Product 927 for Samples 156 to 185

44	60	54	74	61	40	50	50	42	45	46	37	51	27	15
27	24	33	15	15	35	27	30	35	35	29	40	40	25	64

- Use Part One of the following worksheet to find the Effective Cost of Production for the period covered by Samples 156 to 185. The average value is 39.0 *ppt*, and the standard deviation statistic is $s = 14.8$ *ppt*.

- Place these 30 values on an *XmR* chart and use Part Three of the following worksheet to find the Predictable Cost of Production. The average moving range is 10.55 *ppt*. To compute the *Hypothetical DNS* value assume that as you learn how to operate this process predictably you will also be able to move the process average down to 25 *ppt*.

Figure 6.6 shows that this process has the potential to operate with an Effective Cost of Production of 114%. However, the only way they will be able to operate at this level will be to find and remove the assignable causes that led to the deterioration following Sample 126.

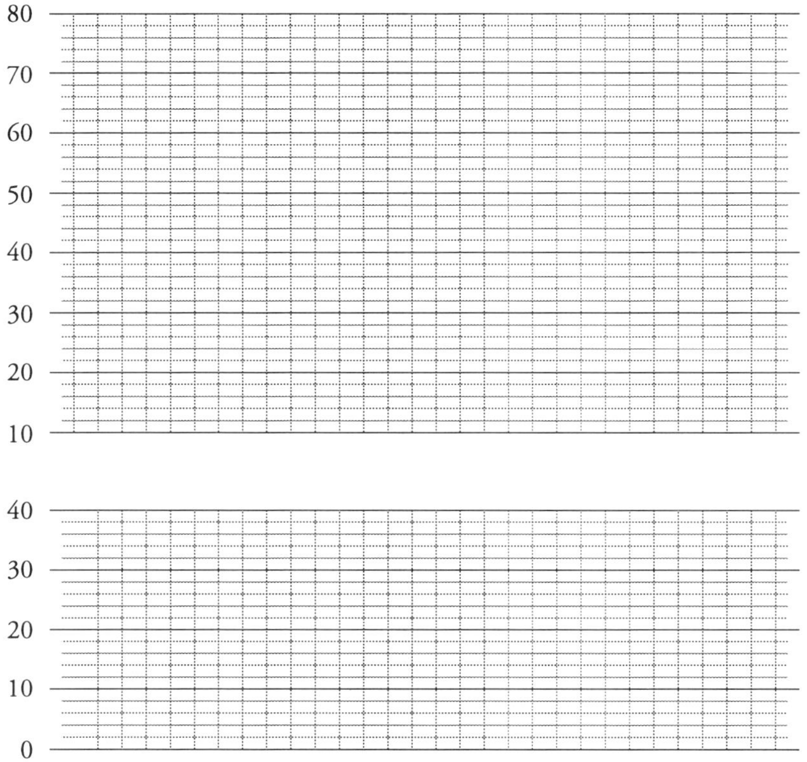
80
70
60
50
40
30
20
10
40
30
20
10
0

Worksheet for Target at Boundary

Part One: Characterizing Past Performance

1. Identify a critical product characteristic:

Product Identification: ______________________

Critical Characteristic Evaluated: ______________________

Specification Limit = SL = __________

Boundary = Target = __________

2. Compute a *scale factor* using a global measure of dispersion:

Number of values used = N = ______

scale factor: s = __________ or R/d_2 = __________

3. Characterize the Elbow Room:

$$\textit{Elbow Room} = \frac{|\ target - SL\ |}{scale\ factor} = \_\_\_\_\_\_ = \_\_\_\_\_\_$$

4. Characterize the Distance to (Nearer) Specification:

average value for the N data = __________

$$DNS = \frac{|\ average - SL\ |}{scale\ factor} = \_\_\_\_\_\_ = \_\_\_\_\_\_$$

(DNS values are defined to be negative whenever the average falls on the wrong side of the specification limit.)

5. Find the Effective Cost of Production:

Is nonconforming product scrapped or reworked?

If it is reworked, then what is the cost of rework as a proportion of the nominal cost of production? 100% 80% 67% 50% 33% 20% 10%

Choose the appropriate table from Tables 30 to 37, use the Elbow Room and DNS values found above, and read off your **Effective Cost of Production**: [ ]

Worksheet for Target at Boundary

page 2 of 5

Part Two: Finding the Optimum Aim Point (optional)

(If the nominal cost of production increases as you move the process average closer to the target value, then it will be of interest to determine the optimum aim point for your process:)

6. For the *DNS* values shown, find the aim points:

aim point = specification ± [DNS x scale factor]

DNS	Aim Point	Nominal Cost	Effective Cost	Actual Cost
0.3	______	______	______	______
0.6	______	______	______	______
0.9	______	______	______	______
1.2	______	______	______	______
1.5	______	______	______	______
1.8	______	______	______	______
2.1	______	______	______	______
2.4	______	______	______	______
2.7	______	______	______	______
3.0	______	______	______	______
3.3	______	______	______	______
3.6	______	______	______	______
3.9	______	______	______	______
4.2	______	______	______	______
4.5	______	______	______	______
4.8	______	______	______	______
5.1	______	______	______	______
5.4	______	______	______	______
5.7	______	______	______	______
6.0	______	______	______	______

Worksheet for Target at Boundary

page 3 of 5

DNS	Aim Point	Nominal Cost	Effective Cost	Actual Cost
6.3	______	______	______	______
6.6	______	______	______	______
6.9	______	______	______	______
7.2	______	______	______	______
7.5	______	______	______	______
7.8	______	______	______	______
8.1	______	______	______	______
8.4	______	______	______	______
8.7	______	______	______	______
9.0	______	______	______	______
9.3	______	______	______	______
9.6	______	______	______	______
9.9	______	______	______	______
10.2	______	______	______	______
10.5	______	______	______	______
10.8	______	______	______	______
11.1	______	______	______	______
11.4	______	______	______	______
11.7	______	______	______	______
12.0	______	______	______	______

7. **For each of the aim points above find the nominal cost of production:**
8. **Use the table from Step 5 to obtain the Effective Costs of Production:**
9. **Multiply the nominal costs by the Effective Costs to find the actual costs of production:**
   Identify the aim point that gives the minimum actual cost.

Optimum aim point = ____________

Worksheet for Target at Boundary

page 4 of 5

Part Three: Potential Payoff for Operating Predictably

10. Place your past performance data on a process behavior chart:

☐ If your process appears to be predictable, skip the remainder of this worksheet. Unless or until your process is changed in some major way, the cost of production in Step 5 is what you should expect from your process in the future. If you improve your process aim you may do as well as shown in Step 9.

☐ If your process appears to be unpredictable, then you will need to compute a *Potential Cost of Production*. As a first step toward this value use the Average Range, or the Average Moving Range, from your process behavior chart and the appropriate bias correction factor from Table 38 to compute a *predictable scale factor:*

subgroup size, n = ______ (use $n = 2$ for XmR chart)

Average Range = $\bar{R}$ = ____________

predictable scale factor = $\dfrac{\bar{R}}{d_2}$ = ____________ = ____________

11. Evaluate the benefits of operating predictably at current average:
Use the predictable scale factor above to compute hypothetical Elbow Room and *DNS* values:

$$\textit{Hypothetical Elbow Room} = \frac{|\ target - SL\ |}{\textit{predictable scale factor}} = \underline{\qquad\qquad}$$

$$\textit{Hypothetical DNS} = \frac{|\ average - SL\ |}{\textit{predictable scale factor}} = \underline{\qquad\qquad}$$

Use these values with the table from Step 5 to find a

Potential Cost of Production = ____________

Worksheet for Target at Boundary

page 5 of 5

12. Evaluate the benefits of operating predictably at optimum aim:
If Part Two of this worksheet was appropriate, then you may wish to use the predictable scale factor and the optimum aim point to compute a new hypothetical DNS value:

$$\textit{New Hypothetical DNS} = \frac{|\ \textit{optimum aim} - SL\ |}{\textit{predictable scale factor}} = \underline{\hspace{4cm}}$$

Use this value with the Hypothetical Elbow Room from Step 11, along with the table from Step 5 to find a

Potential Cost of Production at Optimum Aim = ______________

Chapter Seven

Process Specifications

As noted earlier, there is a distinct difference between specifications which are used to take action on the product and specifications that are used to take action on the process. This chapter will consider how to take appropriate actions based on process specifications.

7.1 Count-Based Process Specifications

What happens when it is impossible to measure the items? What happens when each item may be characterized as acceptable or unacceptable, but no measurement can be made to quantify each item relative to some specification limit? We may still count the good items and the bad items, and compute the fraction of nonconforming product, but there is no underlying scale with a dividing line between acceptable product and unacceptable product. And this type of scale is part of most of the graphs in the preceding chapters. When this scale is missing, the specifications are generally written for the process rather than the product: the specifications define a maximum percentage, or an average percentage, of nonconforming product to be found in the product stream.

When the data are based on counts, specifications will apply to the fraction nonconforming in the product stream, and therefore will be process specifications rather than product specifications. Of course there may need to be some definition of what to count as conforming and nonconforming, but these definitions will be always be combined with a process specification.

So what do you do with process specifications? Consider the following data for stamped parts. Because of the complex shape of the part it was impossible to determine if the part would fit into the assembly by physical measurements, therefore they shipped all of the parts produced and the customer would cull the parts that did not fit. Periodically he would file a claim for those that he culled. The data from 22 reports are shown. These values consist of the number of bad parts, the number of parts used, and the percentage of bad parts. Since the parts were shipped in containers of 40 pieces each, the numbers used are all multiples of 40. The specification is 2.5 percent nonfunctional.

Report	No. Bad	No. Used	Percent Bad	Report	No. Bad	No. Used	Percent Bad
-1-	4	400	1.0	-12-	7	320	2.2
-2-	4	120	3.3	-13-	6	200	3.0
-3-	5	200	2.5	-14-	2	160	1.3
-4-	8	280	2.9	-15-	2	200	1.0
-5-	2	200	1.0	-16-	5	280	1.8
-6-	3	200	1.5	-17-	5	360	1.4
-7-	2	200	1.0	-18-	5	200	2.5
-8-	2	160	1.3	-19-	4	200	2.0
-9-	5	200	2.5	-20-	4	160	2.5
-10-	9	200	4.5	-21-	8	200	4.0
-11-	7	320	2.2	-22-	7	200	3.5

So how can you characterize this process? Some days are better than others? If you are going to use process specifications, then you need to characterize your process, and the premier technique for doing that is a process behavior chart.

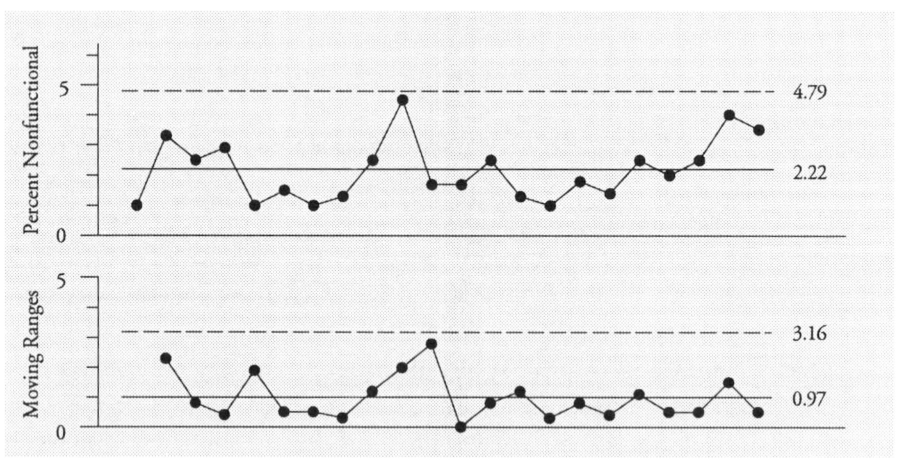

Figure 7.1: ***XmR*** **Chart for Percent Nonfunctional Stampings**

These percentages are placed on an *XmR* chart in Figure 7.1. There you can see that this process is predictable and is averaging about 2.2 percent nonfunctional. In any one report the fraction nonfunctional might vary up to 5%, but the long-term average is close to 2% nonfunctional.

So should they be concerned about report number 10 which had 4.5% nonfunctional? Does 4.5% nonfunctional signify a change in this process? Did things take a turn for the worse? The process behavior chart tells us that the 4.5% is within the bounds of routine variation and should not be interpreted, by itself, as a signal of any change in this process. Subsequent points on the chart bear out this interpretation.

But what does the specification of no more than 2.5% nonfunctional mean? If the specification is "no more than 2.5% nonfunctional on the average," then this process is capable of meeting this requirement. Since this process is predictable we can say that, unless it changes for the worse, it will average about 2.2% nonfunctional, and will therefore meet this specification.

If the specification is "no more than 2.5% nonfunctional in any one period of time," then this process is not capable of meeting this requirement. Therefore, it is critical to clarify exactly what a process specification means. Is it defining the average level of nonconformity, or an absolute maximum?

Finally, with process specifications, the process behavior chart for the product stream defines the actual capability of a predictable process, or the potential capability of an unpredictable process. The central line defines what to expect over the long term, and the limits show the amount of short-term, routine variation about that long-term average that you should expect in practice.

7.2 Measurement-Based Process Specifications

When items are characterized as good or bad, and items are counted, the specification applies to the product stream. But this characterization can also occur with measurement based data. While each item may be measured, and while there may be a specification limit that each item will either meet or fail to meet, if the nonconforming items are not sorted out for separate disposition then the specification becomes a process specification even though it is stated in terms of a measurable product characteristic.

In order for a specification to be considered as a *product* specification it will need to be used to decide which items to rework or scrap and which items to ship. Whenever a specification on a product characteristic is used as a classification device in order to determine the percentage nonconforming, and this information is used to make adjustments to the process rather than taking action on the product, then the specification is a *process* specification.

One difference between count-based process specifications and measurement-based process specifications is what you put on your process behavior chart. With count-based process specifications you will plot the percentage conforming or nonconforming. With measurement-based process specifications you can use the measurements to create a process behavior chart for the product characteristic.

An example of a measurement-based process specification is net weight. The net weight applies to each item produced, but the underweight containers are rarely removed from the product stream. Therefore a net weight is not a guarantee that each item contains so much product, but rather a guarantee that a certain percentage of the containers will contain so much or more of the product. If this percentage drops below the threshold, then action is taken on the process to increase the fill weights.

In this case the excess cost function can be thought of as a step function. Each item either does meet the net weight or it does not meet it. There is no degradation of functionality as a container approaches the net weight. However, because the actions are taken on the process, the concept of the Effective Cost of Production does not apply. The cost of production is simply the nominal cost for the average fill weight.

Of course, it is the difference between the average fill weight and the net weight that is of concern to the producer. This difference is commonly called the "give-away." You will want to meet the net weight requirement while minimizing the give-away. This requires that you have the ability to operate your process predictably and on-target.

7.3 Characterizing Past Performance

When working with a measurement-based process specification you can use the measurements to characterize past performance. As an example consider the 50 fill weights shown below. These values were obtained over an extended period of time. The average fill value is 28.08 ounces and the global standard deviation statistic is s = 0.2614 ounces per standard deviation.

27.7	27.7	27.9	28.1	28.1	28.3	28.2	28.3	28.1	28.3
28.1	28.0	28.0	28.1	28.2	28.4	28.4	28.5	28.6	28.6
28.5	28.5	28.3	28.3	28.4	28.3	28.3	28.1	28.2	28.1
28.0	28.1	28.2	28.1	27.9	28.0	27.9	27.7	27.9	27.8
28.0	27.9	27.8	27.8	27.8	27.9	27.8	27.6	27.7	27.7

The net weight for this product is 26.8 ounces and the requirement is that 99 percent of the containers shall meet or exceed the net weight. The histogram shows that all 50 of these containers exceeded the net weight. The average fill weight of 28.08 ounces represents an overfill of 1.28 ounces, or 4.8 percent give-away. Also, as shown in Figure 7.2, the excess cost function is a step function, with no penalty for being close to the specification.

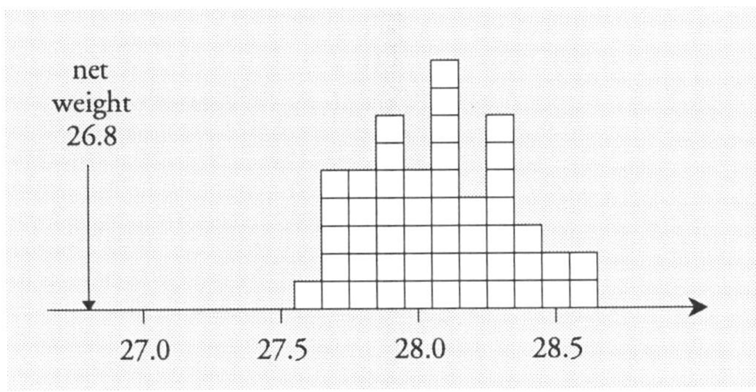

Figure 7.2: Histogram of 50 Fill Weights

The Distance to Nearer Specification is:

$$DNS = \frac{28.08\ oz. - 26.8\ oz.}{0.2614\ oz.\ per\ std.\ dev.} = 4.90\ standard\ deviations$$

This *DNS* value would suggest that the process aim could safely be reduced. For 99 percent conformance the past performance data would suggest an aim point of:

$$aim\ point = 26.8\ oz. + [\ 2.5\ std.\ dev. \times 0.261\ oz.\ per\ std.\ dev.\ ] = 27.45\ oz.$$

However, this aim point assumes that the past can be used as a guide to the future. And this assumption has not yet been examined.

7.4 Characterizing Process Behavior

When a process displays predictable behavior it can be thought of as operating in a steady state, with a fixed amount of routine variation. In this case, any reasonable sample drawn from the process will characterize the whole process, and you can safely assume that the items that you measure are like the items that you did not measure. In other words, the process behavior chart will examine the data stream for homogeneity. If the data stream is found to be reasonably homogeneous, then any portion of that data stream will characterize the whole data stream.

However, when a process displays unpredictable behavior it must be thought of as being subject to assignable causes of exceptional variation. These assignable causes will disrupt the homogeneity of the data stream, making it difficult to use any one portion of the data stream to represent the whole. When this happens you cannot safely extrapolate from the product you have measured to the product you have not measured, nor can you safely extrapolate from the past to the future.

And yet the essence of doing business requires prediction.

Fortunately, the process behavior chart will not only characterize the process behavior, but it will also define the hypothetical capability of an unpredictable process. Returning to the 50 fill weights, the *XmR* chart for these values is shown in Figure 7.3. There we see that this process is behaving unpredictably.

This process does not have a single process average, but many. And when this is the case the important question is not, "What is the process average?," but "Why is the process average changing?" Until you can answer the second question, the first question is meaningless. Moreover, until you have done something to keep the process average from changing in the future, you will never know if the containers you measure will truly represent the containers you do not measure, and you will not know where to set the process average to minimize the give-away. Moreover, with an unpredictable process, setting the process aim based on historical averages is equivalent to optimizing for the past, rather than dealing with the present or the future.

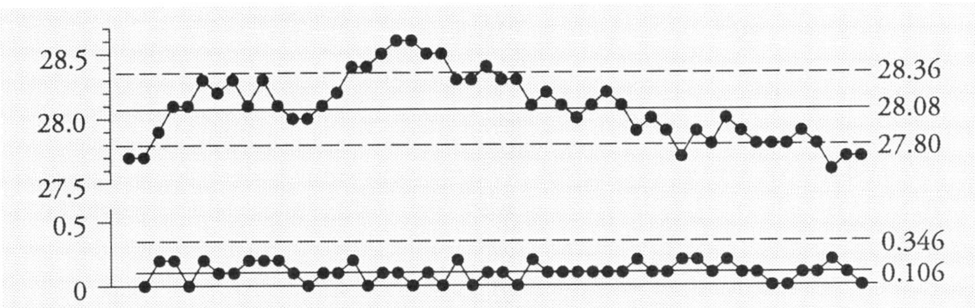

Figure 7.3: *XmR* Chart for 50 Fill Weights

So now that we know that this process is behaving unpredictably, we know better than to use Figure 7.2 as a prediction of what to expect in the future. But what can we say about this process? As noted earlier, the process behavior chart does define the hypothetical capability of this process. Using the average moving range we find a predictable scale factor to be:

$$\textit{predictable scale factor} = \frac{\bar{R}}{d_2} = \frac{0.106}{1.128} = 0.094 \textit{ ounces per std. dev.}$$

Given that we want 99 percent conformance we might choose a process aim point that is about 2.5 standard deviations above the net weight. In this case such an aim point would be:

$$\textit{potential aim point} = 26.8 \textit{ oz.} + [\, 2.5 \textit{ s. d.} \times 0.094 \textit{ oz. per s. d.}\,] = 27.03 \textit{ oz.}$$

Thus, the potential savings for operating this process predictably at the optimum aim is a reduction in the aim point from the current 28.08 oz. to 27.03 oz. This would reduce the give-away from 4.8 percent to 0.9 percent, a savings on fill material of 3.9 percent.

Notice that you can directly convert numbers like these into costs and savings for various courses of action. You do not need to compute capability indexes first. Since process specifications focus directly upon the process, rather than some product characteristic, the conversion that was accomplished by the capability indexes and related measures is no longer necessary. You can directly evaluate the economic consequences of a process improvement. In fact, when you have process specifications instead of product specifications, the essence of process evaluation is the characterization of the process itself, and the process behavior chart is the primary tool for this characterization. Therefore, no worksheet is given for use with process specifications. Place your data on an appropriate process behavior chart, learn what it is telling you about your process, evaluate the potential savings of different courses of action, and take appropriate action.

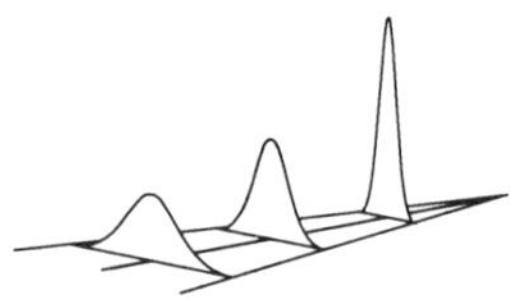

Tables 1 to 20
The Effective Cost of Production for Two-Sided Specifications

Table 1: The Effective Cost of Production when *all* nonconforming units are scrapped

C_{pk} \ C_p values	0.10	0.20	0.30	0.40	0.50	0.60	0.70	0.80	0.90	1.00
–0.50	20.83	16.19	15.54	15.51	15.56	15.61	15.65	15.68	15.71	15.73
–0.40	13.02	9.74	9.22	9.19	9.24	9.29	9.33	9.37	9.40	9.42
–0.30	8.89	6.38	5.93	5.90	5.94	5.99	6.04	6.07	6.11	6.13
–0.20	6.64	4.54	4.13	4.08	4.11	4.16	4.21	4.25	4.28	4.31
–0.10	5.40	3.51	3.09	3.02	3.04	3.09	3.13	3.17	3.21	3.24
0.00	4.77	2.93	2.48	2.38	2.38	2.42	2.47	2.51	2.55	2.58
0.10	4.58	2.63	2.13	1.98	1.97	1.99	2.03	2.08	2.11	2.15
0.20	-	2.54	1.94	1.74	1.69	1.71	1.74	1.78	1.82	1.85
0.30	-	-	1.88	1.61	1.52	1.51	1.54	1.57	1.61	1.64
0.40	-	-	-	1.57	1.43	1.39	1.39	1.42	1.45	1.48
0.50	-	-	-	-	1.40	1.32	1.29	1.31	1.33	1.36
0.60	-	-	-	-	-	1.29	1.24	1.23	1.25	1.27
0.70	-	-	-	-	-	-	1.22	1.19	1.18	1.20
0.80	-	-	-	-	-	-	-	1.17	1.15	1.15
0.90	-	-	-	-	-	-	-	-	1.14	1.12
1.00	-	-	-	-	-	-	-	-	-	1.11

Blank spaces correspond to impossible combinations of C_p and C_{pk}.

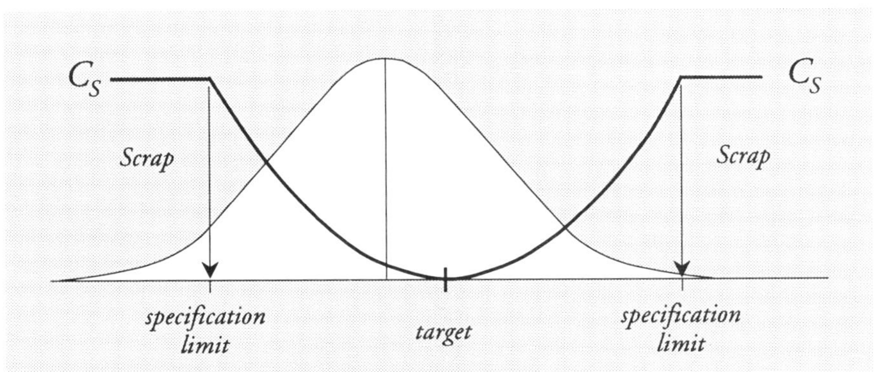

Table 1: Effective Cost of Production When All Nonconforming Items Are Scrapped

Table 1: The Effective Cost of Production when *all* nonconforming units are scrapped

	C_p values									
C_{pk}	***1.10***	***1.20***	***1.30***	***1.40***	***1.50***	***1.60***	***1.70***	***1.80***	***1.90***	***2.00***
–0.50	15.75	15.77	15.78	15.80	15.81	15.82	15.83	15.83	15.84	15.85
–0.40	9.44	9.46	9.48	9.49	9.51	9.52	9.53	9.53	9.54	9.55
–0.30	6.16	6.18	6.19	6.21	6.22	6.23	6.24	6.25	6.26	6.27
–0.20	4.33	4.36	4.38	4.39	4.41	4.42	4.43	4.44	4.45	4.46
–0.10	3.27	3.29	3.31	3.33	3.35	3.36	3.37	3.39	3.40	3.41
0.00	2.61	2.63	2.66	2.68	2.70	2.71	2.73	2.74	2.75	2.76
0.10	2.18	2.21	2.23	2.25	2.27	2.29	2.31	2.32	2.34	2.35
0.20	1.89	1.92	1.94	1.97	1.99	2.01	2.03	2.04	2.06	2.07
0.30	1.68	1.71	1.74	1.76	1.79	1.81	1.83	1.84	1.86	1.87
0.40	1.52	1.55	1.58	1.61	1.63	1.66	1.68	1.70	1.71	1.73
0.50	1.40	1.43	1.46	1.49	1.51	1.54	1.56	1.58	1.60	1.62
0.60	1.30	1.33	1.36	1.39	1.42	1.44	1.47	1.49	1.51	1.53
0.70	1.22	1.25	1.28	1.31	1.34	1.36	1.39	1.41	1.43	1.46
0.80	1.17	1.19	1.21	1.24	1.27	1.29	1.32	1.34	1.37	1.39
0.90	1.12	1.14	1.16	1.18	1.21	1.24	1.26	1.28	1.31	1.33
1.00	1.10	1.10	1.12	1.14	1.16	1.18	1.21	1.23	1.26	1.28
1.10	1.09	1.08	1.09	1.10	1.12	1.14	1.16	1.19	1.21	1.23
1.20	-	1.08	1.07	1.08	1.09	1.11	1.12	1.15	1.17	1.19
1.30	-	-	1.07	1.06	1.07	1.08	1.09	1.11	1.13	1.15
1.40	-	-	-	1.06	1.05	1.06	1.07	1.08	1.10	1.12
1.50	-	-	-	-	1.05	1.05	1.05	1.06	1.08	1.09
1.60	-	-	-	-	-	1.04	1.04	1.05	1.06	1.07
1.70	-	-	-	-	-	-	1.04	1.04	1.04	1.05
1.80	-	-	-	-	-	-	-	1.03	1.03	1.04
1.90	-	-	-	-	-	-	-	-	1.03	1.03
2.00	-	-	-	-	-	-	-	-	-	1.03

Table 1: The Effective Cost of Production when *all* nonconforming units are scrapped

C_{pk}	C_p values 2.5	3.0	4.0	5.0	6.0	8.0	12.0	20.0
–0.50	15.91	15.93	15.95	15.97	15.98	15.99	16.00	16.01
–0.40	9.59	9.61	9.64	9.65	9.66	9.68	9.69	9.70
–0.30	6.31	6.33	6.36	6.38	6.39	6.40	6.42	6.43
–0.20	4.50	4.52	4.56	4.57	4.59	4.60	4.62	4.63
–0.10	3.45	3.48	3.51	3.53	3.55	3.56	3.58	3.60
0.00	2.80	2.83	2.87	2.90	2.92	2.94	2.96	2.98
0.10	2.40	2.43	2.48	2.50	2.52	2.55	2.57	2.59
0.20	2.12	2.16	2.21	2.24	2.27	2.29	2.32	2.34
0.30	1.94	1.98	2.04	2.07	2.10	2.13	2.16	2.19
0.40	1.80	1.85	1.91	1.95	1.98	2.02	2.05	2.08
0.50	1.70	1.75	1.82	1.87	1.90	1.94	1.98	2.02
0.60	1.61	1.67	1.75	1.81	1.84	1.89	1.94	1.98
0.70	1.54	1.61	1.70	1.76	1.80	1.85	1.90	1.95
0.80	1.48	1.55	1.65	1.72	1.76	1.82	1.88	1.93
0.90	1.43	1.50	1.61	1.68	1.73	1.79	1.86	1.92
1.00	1.38	1.46	1.57	1.65	1.70	1.77	1.84	1.90
1.10	1.33	1.41	1.53	1.61	1.67	1.75	1.83	1.89
1.20	1.29	1.37	1.50	1.58	1.64	1.72	1.81	1.88
1.30	1.25	1.33	1.46	1.55	1.62	1.70	1.80	1.87
1.40	1.21	1.30	1.43	1.52	1.59	1.68	1.78	1.87
1.50	1.18	1.26	1.40	1.49	1.57	1.66	1.77	1.86
1.60	1.15	1.23	1.37	1.47	1.54	1.64	1.75	1.85
1.70	1.12	1.20	1.34	1.44	1.52	1.62	1.74	1.84
1.80	1.10	1.17	1.31	1.41	1.49	1.60	1.72	1.83
1.90	1.08	1.15	1.28	1.39	1.47	1.58	1.71	1.82
2.00	1.06	1.12	1.26	1.36	1.45	1.56	1.70	1.81
2.5	1.02	1.04	1.15	1.25	1.34	1.47	1.63	1.77
3.0	-	1.01	1.07	1.16	1.25	1.39	1.56	1.72
4.0	-	-	1.01	1.04	1.11	1.25	1.45	1.64
5.0	-	-	-	1.00	1.03	1.14	1.34	1.56
6.0	-	-	-	-	1.00	1.06	1.25	1.49
8.0	-	-	-	-	-	1.00	1.11	1.36
12.0	-	-	-	-	-	-	1.00	1.16
20.0	-	-	-	-	-	-	-	1.00

Table 2: The Effective Cost of Production when *all* nonconforming units are reworked

C_{pk}	C_p values 0.10	0.20	0.30	0.40	0.50	0.60	0.70	0.80	0.90	1.00
1.00	-	-	-	-	-	-	-	-	-	1.11
0.90	-	-	-	-	-	-	-	-	1.14	1.12
0.80	-	-	-	-	-	-	-	1.17	1.15	1.15
0.70	-	-	-	-	-	-	1.21	1.18	1.18	1.20
0.60	-	-	-	-	-	1.27	1.23	1.22	1.24	1.26
0.50	-	-	-	-	1.35	1.29	1.27	1.29	1.31	1.34
0.40	-	-	-	1.44	1.36	1.34	1.35	1.37	1.40	1.43
0.30	-	-	1.56	1.46	1.42	1.42	1.44	1.46	1.49	1.52
0.20	-	1.69	1.57	1.51	1.50	1.51	1.54	1.57	1.59	1.62
0.10	1.84	1.71	1.62	1.59	1.59	1.61	1.64	1.66	1.69	1.71
0.00	1.85	1.74	1.69	1.68	1.69	1.71	1.73	1.75	1.77	1.79
–0.10	1.87	1.79	1.76	1.76	1.78	1.80	1.81	1.83	1.84	1.86
–0.20	1.89	1.84	1.83	1.84	1.85	1.87	1.88	1.89	1.90	1.91
–0.30	1.92	1.89	1.89	1.90	1.91	1.92	1.93	1.93	1.94	1.94
–0.40	1.95	1.93	1.93	1.94	1.95	1.95	1.96	1.96	1.96	1.97
–0.50	1.97	1.96	1.96	1.97	1.97	1.97	1.98	1.98	1.98	1.98

Blank spaces correspond to impossible combinations of C_p and C_{pk}.

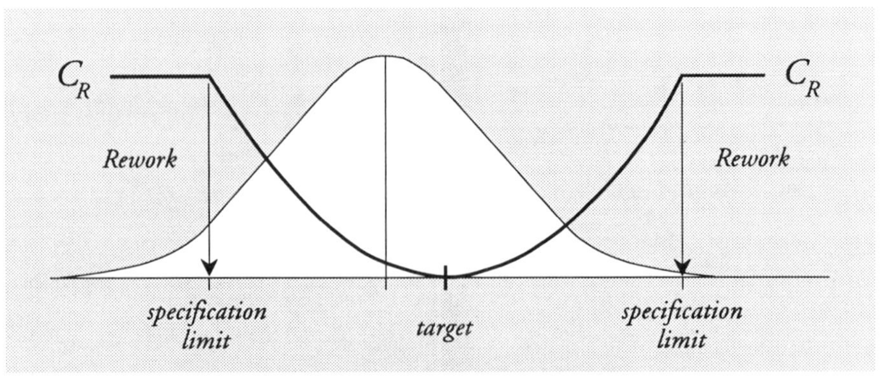

Table 2: Effective Cost of Production When All Nonconforming Items Are Reworked

Table 2: The Effective Cost of Production when *all* nonconforming units are reworked

C_{pk}	C_p values: ***1.10***	***1.20***	***1.30***	***1.40***	***1.50***	***1.60***	***1.70***	***1.80***	***1.90***	***2.00***
2.00	-	-	-	-	-	-	-	-	-	1.03
1.90	-	-	-	-	-	-	-	-	1.03	1.03
1.80	-	-	-	-	-	-	-	1.03	1.03	1.04
1.70	-	-	-	-	-	-	1.04	1.04	1.04	1.05
1.60	-	-	-	-	-	1.04	1.04	1.05	1.06	1.07
1.50	-	-	-	-	1.05	1.05	1.05	1.06	1.08	1.09
1.40	-	-	-	1.06	1.05	1.06	1.07	1.08	1.10	1.12
1.30	-	-	1.07	1.06	1.07	1.08	1.09	1.11	1.13	1.15
1.20	-	1.08	1.07	1.08	1.09	1.11	1.12	1.15	1.17	1.19
1.10	1.09	1.08	1.09	1.10	1.12	1.14	1.16	1.19	1.21	1.23
1.00	1.10	1.10	1.12	1.14	1.16	1.18	1.21	1.23	1.26	1.28
0.90	1.12	1.14	1.16	1.18	1.21	1.23	1.26	1.28	1.31	1.33
0.80	1.16	1.19	1.21	1.24	1.27	1.29	1.32	1.34	1.36	1.39
0.70	1.22	1.25	1.28	1.30	1.33	1.36	1.38	1.41	1.43	1.45
0.60	1.29	1.32	1.35	1.38	1.40	1.43	1.45	1.47	1.49	1.51
0.50	1.37	1.40	1.43	1.45	1.48	1.50	1.52	1.55	1.56	1.58
0.40	1.46	1.49	1.51	1.54	1.56	1.58	1.60	1.62	1.63	1.65
0.30	1.55	1.58	1.60	1.62	1.64	1.66	1.67	1.69	1.70	1.71
0.20	1.64	1.66	1.68	1.70	1.72	1.73	1.74	1.76	1.77	1.78
0.10	1.73	1.75	1.76	1.77	1.79	1.80	1.81	1.82	1.82	1.83
0.00	1.80	1.82	1.83	1.84	1.85	1.85	1.86	1.87	1.87	1.88
–0.10	1.87	1.87	1.88	1.89	1.90	1.90	1.91	1.91	1.91	1.92
–0.20	1.91	1.92	1.92	1.93	1.93	1.94	1.94	1.95	1.95	1.95
–0.30	1.95	1.95	1.95	1.96	1.96	1.96	1.97	1.97	1.97	1.97
–0.40	1.97	1.97	1.97	1.98	1.98	1.98	1.98	1.98	1.98	1.98
–0.50	1.98	1.99	1.99	1.99	1.99	1.99	1.99	1.99	1.99	1.99

Table 2: The Effective Cost of Production when *all* nonconforming units are reworked

	C_p values							
C_{pk}	**2.5**	**3.0**	**4.0**	**5.0**	**6.0**	**8.0**	**12.0**	**20.0**
20.0	-	-	-	-	-	-	-	1.00
12.0	-	-	-	-	-	-	1.00	1.16
8.0	-	-	-	-	-	1.00	1.11	1.36
6.0	-	-	-	-	1.00	1.06	1.25	1.49
5.0	-	-	-	1.00	1.03	1.14	1.34	1.56
4.0	-	-	1.01	1.04	1.11	1.25	1.45	1.64
3.0	-	1.01	1.07	1.16	1.25	1.39	1.56	1.72
2.5	1.02	1.04	1.15	1.25	1.34	1.47	1.63	1.77
2.00	1.06	1.12	1.26	1.36	1.45	1.56	1.70	1.81
1.90	1.08	1.15	1.28	1.39	1.47	1.58	1.71	1.82
1.80	1.10	1.17	1.31	1.41	1.49	1.60	1.72	1.83
1.70	1.12	1.20	1.34	1.44	1.52	1.62	1.74	1.84
1.60	1.15	1.23	1.37	1.47	1.54	1.64	1.75	1.85
1.50	1.18	1.26	1.40	1.49	1.57	1.66	1.77	1.86
1.40	1.21	1.30	1.43	1.52	1.59	1.68	1.78	1.87
1.30	1.25	1.33	1.46	1.55	1.62	1.70	1.80	1.87
1.20	1.29	1.37	1.50	1.58	1.64	1.72	1.81	1.88
1.10	1.33	1.41	1.53	1.61	1.67	1.75	1.83	1.89
1.00	1.38	1.46	1.57	1.64	1.70	1.77	1.84	1.90
0.90	1.43	1.50	1.61	1.68	1.73	1.79	1.86	1.91
0.80	1.48	1.55	1.65	1.71	1.75	1.81	1.87	1.92
0.70	1.53	1.60	1.69	1.74	1.78	1.83	1.89	1.93
0.60	1.59	1.65	1.73	1.78	1.81	1.86	1.90	1.94
0.50	1.65	1.70	1.77	1.81	1.84	1.88	1.92	1.95
0.40	1.71	1.75	1.81	1.84	1.87	1.90	1.93	1.96
0.30	1.76	1.80	1.85	1.87	1.89	1.92	1.95	1.97
0.20	1.82	1.84	1.88	1.90	1.92	1.94	1.96	1.97
0.10	1.86	1.88	1.91	1.93	1.94	1.95	1.97	1.98
0.00	1.90	1.92	1.94	1.95	1.96	1.97	1.98	1.99
–0.10	1.93	1.94	1.96	1.97	1.97	1.98	1.99	1.99
–0.20	1.96	1.96	1.97	1.98	1.98	1.99	1.99	2.00
–0.30	1.98	1.98	1.98	1.99	1.99	1.99	1.99	2.00
–0.40	1.99	1.99	1.99	1.99	1.99	2.00	2.00	2.00
–0.50	1.99	2.00	2.00	2.00	2.00	2.00	2.00	2.00

Table 3: The Effective Cost of Production when the cost of rework is equal to the cost of scrap and the average is on the SCRAP side of the target

	C_p *values*									
C_{pk}	***0.10***	***0.20***	***0.30***	***0.40***	***0.50***	***0.60***	***0.70***	***0.80***	***0.90***	***1.00***
–0.50	15.51	15.40	15.43	15.50	15.56	15.61	15.65	15.68	15.71	15.73
–0.40	9.25	9.11	9.12	9.18	9.24	9.29	9.33	9.37	9.40	9.42
–0.30	6.02	5.85	5.84	5.88	5.94	5.99	6.04	6.07	6.11	6.13
–0.20	4.26	4.08	4.03	4.06	4.11	4.16	4.21	4.25	4.28	4.31
–0.10	3.27	3.07	2.99	3.00	3.04	3.09	3.13	3.17	3.21	3.24
0.00	2.70	2.48	2.37	2.35	2.38	2.42	2.47	2.51	2.55	2.58
0.10	2.36	2.14	2.00	1.95	1.96	1.99	2.03	2.08	2.11	2.15
0.20	-	1.95	1.79	1.71	1.69	1.71	1.74	1.78	1.82	1.85
0.30	-	-	1.68	1.56	1.51	1.51	1.53	1.57	1.61	1.64
0.40	-	-	-	1.50	1.41	1.38	1.39	1.42	1.45	1.48
0.50	-	-	-	-	1.37	1.31	1.29	1.31	1.33	1.36
0.60	-	-	-	-	-	1.28	1.24	1.23	1.25	1.27
0.70	-	-	-	-	-	-	1.22	1.19	1.18	1.20
0.80	-	-	-	-	-	-	-	1.17	1.15	1.15
0.90	-	-	-	-	-	-	-	-	1.14	1.12
1.00	-	-	-	-	-	-	-	-	-	1.11

Blank spaces correspond to impossible combinations of C_p and C_{pk}.

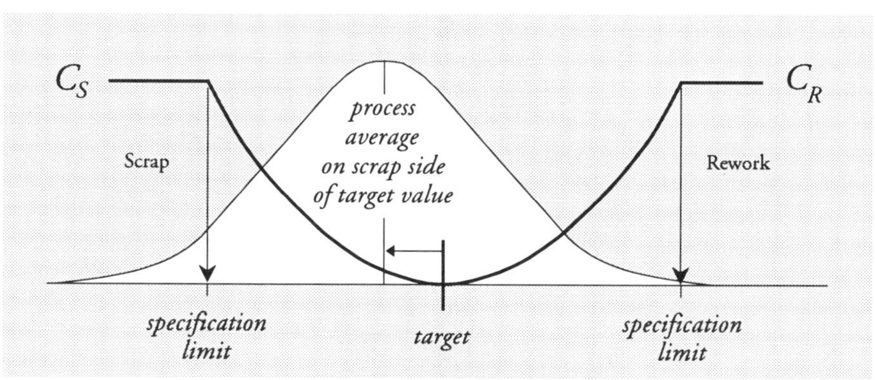

Table 3: Effective Cost of Production
When Cost of Rework is Equal to the Cost of Scrap
and the Average is on the Scrap Side of the Target Value

Table 3: The Effective Cost of Production when the cost of rework is equal to the cost of scrap and the average is on the SCRAP side of the target

	C_p values									
C_{pk}	***1.10***	***1.20***	***1.30***	***1.40***	***1.50***	***1.60***	***1.70***	***1.80***	***1.90***	***2.00***
–0.50	15.75	15.77	15.78	15.80	15.81	15.82	15.83	15.83	15.84	15.85
–0.40	9.44	9.46	9.48	9.49	9.51	9.52	9.53	9.53	9.54	9.55
–0.30	6.16	6.18	6.19	6.21	6.22	6.23	6.24	6.25	6.26	6.27
–0.20	4.33	4.36	4.38	4.39	4.41	4.42	4.43	4.44	4.45	4.46
–0.10	3.27	3.29	3.31	3.33	3.35	3.36	3.37	3.39	3.40	3.41
0.00	2.61	2.63	2.66	2.68	2.70	2.71	2.73	2.74	2.75	2.76
0.10	2.18	2.21	2.23	2.25	2.27	2.29	2.31	2.32	2.34	2.35
0.20	1.89	1.92	1.94	1.97	1.99	2.01	2.03	2.04	2.06	2.07
0.30	1.68	1.71	1.74	1.76	1.79	1.81	1.83	1.84	1.86	1.87
0.40	1.52	1.55	1.58	1.61	1.63	1.66	1.68	1.70	1.71	1.73
0.50	1.40	1.43	1.46	1.49	1.51	1.54	1.56	1.58	1.60	1.62
0.60	1.30	1.33	1.36	1.39	1.42	1.44	1.47	1.49	1.51	1.53
0.70	1.22	1.25	1.28	1.31	1.34	1.36	1.39	1.41	1.43	1.46
0.80	1.17	1.19	1.21	1.24	1.27	1.29	1.32	1.34	1.37	1.39
0.90	1.12	1.14	1.16	1.18	1.21	1.24	1.26	1.28	1.31	1.33
1.00	1.10	1.10	1.12	1.14	1.16	1.18	1.21	1.23	1.26	1.28
1.10	1.09	1.08	1.09	1.10	1.12	1.14	1.16	1.19	1.21	1.23
1.20	-	1.08	1.07	1.08	1.09	1.11	1.12	1.15	1.17	1.19
1.30	-	-	1.07	1.06	1.07	1.08	1.09	1.11	1.13	1.15
1.40	-	-	-	1.06	1.05	1.06	1.07	1.08	1.10	1.12
1.50	-	-	-	-	1.05	1.05	1.05	1.06	1.08	1.09
1.60	-	-	-	-	-	1.04	1.04	1.05	1.06	1.07
1.70	-	-	-	-	-	-	1.04	1.04	1.04	1.05
1.80	-	-	-	-	-	-	-	1.03	1.03	1.04
1.90	-	-	-	-	-	-	-	-	1.03	1.03
2.00	-	-	-	-	-	-	-	-	-	1.03

When $C_{pk} = C_p$ the average is at the target.

Table 3: The Effective Cost of Production when the cost of rework is equal to the cost of scrap and the average is on the SCRAP side of the target

	C_p values							
C_{pk}	**2.5**	**3.0**	**4.0**	**5.0**	**6.0**	**8.0**	**12.0**	**20.0**
–0.50	15.91	15.93	15.95	15.97	15.98	15.99	16.00	16.01
–0.40	9.59	9.61	9.64	9.65	9.66	9.68	9.69	9.70
–0.30	6.31	6.33	6.36	6.38	6.39	6.40	6.42	6.43
–0.20	4.50	4.52	4.56	4.57	4.59	4.60	4.62	4.63
–0.10	3.45	3.48	3.51	3.53	3.55	3.56	3.58	3.60
0.00	2.81	2.84	2.88	2.90	2.92	2.94	2.96	2.98
0.10	2.40	2.43	2.48	2.50	2.52	2.55	2.57	2.59
0.20	2.12	2.16	2.21	2.24	2.27	2.29	2.32	2.34
0.30	1.94	1.98	2.04	2.07	2.10	2.13	2.16	2.19
0.40	1.80	1.85	1.91	1.95	1.98	2.02	2.05	2.08
0.50	1.70	1.75	1.82	1.87	1.90	1.94	1.98	2.02
0.60	1.61	1.67	1.75	1.81	1.84	1.89	1.94	1.98
0.70	1.54	1.61	1.70	1.76	1.80	1.85	1.90	1.95
0.80	1.48	1.55	1.65	1.72	1.76	1.82	1.88	1.93
0.90	1.43	1.50	1.61	1.68	1.73	1.79	1.86	1.92
1.00	1.38	1.46	1.57	1.65	1.70	1.77	1.84	1.90
1.10	1.33	1.41	1.53	1.61	1.67	1.75	1.83	1.89
1.20	1.29	1.37	1.50	1.58	1.64	1.72	1.81	1.88
1.30	1.25	1.33	1.46	1.55	1.62	1.70	1.80	1.87
1.40	1.21	1.30	1.43	1.52	1.59	1.68	1.78	1.87
1.50	1.18	1.26	1.40	1.49	1.57	1.66	1.77	1.86
1.60	1.15	1.23	1.37	1.47	1.54	1.64	1.75	1.85
1.70	1.12	1.20	1.34	1.44	1.52	1.62	1.74	1.84
1.80	1.10	1.17	1.31	1.41	1.49	1.60	1.72	1.83
1.90	1.08	1.15	1.28	1.39	1.47	1.58	1.71	1.82
2.00	1.06	1.12	1.26	1.36	1.45	1.56	1.70	1.81
2.5	1.02	1.04	1.15	1.25	1.34	1.47	1.63	1.77
3.0	-	1.01	1.07	1.16	1.25	1.39	1.56	1.72
4.0	-	-	1.01	1.04	1.11	1.25	1.45	1.64
5.0	-	-	-	1.00	1.03	1.14	1.34	1.56
6.0	-	-	-	-	1.00	1.06	1.25	1.49
8.0	-	-	-	-	-	1.00	1.11	1.36
12.0	-	-	-	-	-	-	1.00	1.16
20.0	-	-	-	-	-	-	-	1.00

Table 4: The Effective Cost of Production when the cost of rework is equal to the cost of scrap and the average is on the REWORK side of the target

C_{pk}	*C_p values* 0.10	0.20	0.30	0.40	0.50	0.60	0.70	0.80	0.90	1.00
1.00	-	-	-	-	-	-	-	-	-	1.11
0.90	-	-	-	-	-	-	-	-	1.14	1.12
0.80	-	-	-	-	-	-	-	1.17	1.15	1.15
0.70	-	-	-	-	-	-	1.22	1.18	1.18	1.20
0.60	-	-	-	-	-	1.28	1.23	1.22	1.24	1.26
0.50	-	-	-	-	1.37	1.29	1.28	1.29	1.31	1.34
0.40	-	-	-	1.50	1.38	1.34	1.35	1.37	1.40	1.43
0.30	-	-	1.68	1.49	1.43	1.42	1.44	1.46	1.49	1.52
0.20	-	1.95	1.65	1.53	1.50	1.51	1.54	1.57	1.59	1.62
0.10	2.36	1.86	2.66	1.60	1.59	1.61	1.64	1.66	1.69	1.71
0.00	2.17	1.84	1.71	1.68	1.69	1.71	1.73	1.75	1.77	1.79
–0.10	2.06	1.85	1.77	1.77	1.78	1.80	1.81	1.83	1.84	1.86
–0.20	2.01	1.87	1.84	1.84	1.85	1.87	1.88	1.89	1.90	1.91
–0.30	1.99	1.91	1.89	1.90	1.91	1.92	1.93	1.93	1.94	1.94
–0.40	1.98	1.94	1.93	1.94	1.95	1.95	1.96	1.96	1.96	1.97
–0.50	1.99	1.96	1.96	1.97	1.97	1.97	1.98	1.98	1.98	1.98

Blank spaces correspond to impossible combinations of C_p and C_{pk}.

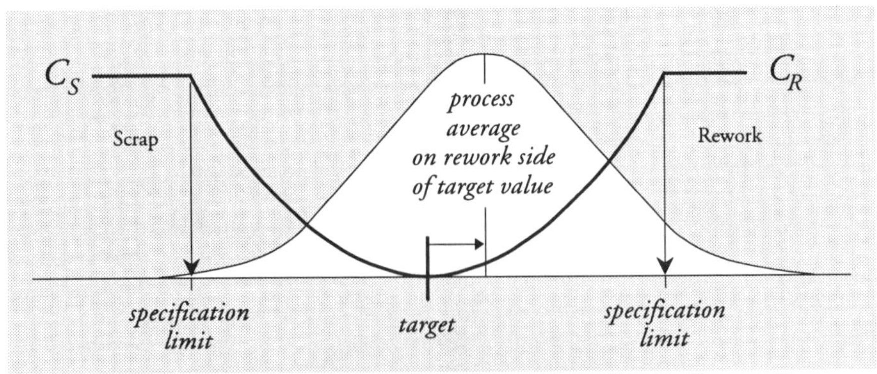

Table 4: Effective Cost of Production
When Cost of Rework is Equal to the Cost of Scrap
and the Average is on the Rework Side of the Target Value

Table 4: The Effective Cost of Production when the cost of rework is equal to the cost of scrap and the average is on the REWORK side of the target

	C_p values									
C_{pk}	***1.10***	***1.20***	***1.30***	***1.40***	***1.50***	***1.60***	***1.70***	***1.80***	***1.90***	***2.00***
2.00	-	-	-	-	-	-	-	-	-	1.03
1.90	-	-	-	-	-	-	-	-	1.03	1.03
1.80	-	-	-	-	-	-	-	1.03	1.03	1.04
1.70	-	-	-	-	-	-	1.04	1.04	1.04	1.05
1.60	-	-	-	-	-	1.04	1.04	1.05	1.06	1.07
1.50	-	-	-	-	1.05	1.05	1.05	1.06	1.08	1.09
1.40	-	-	-	1.06	1.05	1.06	1.07	1.08	1.10	1.12
1.30	-	-	1.07	1.06	1.07	1.08	1.09	1.11	1.13	1.15
1.20	-	1.08	1.07	1.08	1.09	1.11	1.12	1.15	1.17	1.19
1.10	1.09	1.08	1.09	1.10	1.12	1.14	1.16	1.19	1.21	1.23
1.00	1.10	1.10	1.12	1.14	1.16	1.18	1.21	1.23	1.26	1.28
0.90	1.12	1.14	1.16	1.18	1.21	1.23	1.26	1.28	1.31	1.33
0.80	1.16	1.19	1.21	1.24	1.27	1.29	1.32	1.34	1.36	1.39
0.70	1.22	1.25	1.28	1.30	1.33	1.36	1.38	1.41	1.43	1.45
0.60	1.29	1.32	1.35	1.38	1.40	1.43	1.45	1.47	1.49	1.51
0.50	1.37	1.40	1.43	1.45	1.48	1.50	1.52	1.54	1.56	1.58
0.40	1.46	1.49	1.51	1.54	1.56	1.58	1.60	1.62	1.63	1.65
0.30	1.55	1.58	1.60	1.62	1.64	1.66	1.67	1.69	1.70	1.71
0.20	1.64	1.66	1.68	1.70	1.72	1.73	1.74	1.76	1.77	1.78
0.10	1.73	1.75	1.76	1.77	1.79	1.80	1.81	1.82	1.82	1.83
0.00	1.80	1.82	1.83	1.84	1.85	1.85	1.86	1.87	1.87	1.88
–0.10	1.87	1.87	1.88	1.89	1.90	1.90	1.91	1.91	1.91	1.92
–0.20	1.91	1.92	1.92	1.93	1.93	1.94	1.94	1.94	1.95	1.95
–0.30	1.95	1.95	1.95	1.96	1.96	1.96	1.96	1.97	1.97	1.97
–0.40	1.97	1.97	1.97	1.98	1.98	1.98	1.98	1.98	1.98	1.98
–0.50	1.98	1.99	1.99	1.99	1.99	1.99	1.99	1.99	1.99	1.99

When $C_{pk} = C_p$ the average is at the target.

Table 4: The Effective Cost of Production when the cost of rework is equal to the cost of scrap and the average is on the REWORK side of the target

	C_p values							
C_{pk}	2.5	3.0	4.0	5.0	6.0	8.0	12.0	20.0
20.0	-	-	-	-	-	-	-	1.00
12.0	-	-	-	-	-	-	1.00	1.16
8.0	-	-	-	-	-	1.00	1.11	1.36
6.0	-	-	-	-	1.00	1.06	1.25	1.49
5.0	-	-	-	1.00	1.03	1.14	1.34	1.56
4.0	-	-	1.01	1.04	1.11	1.25	1.45	1.64
3.0	-	1.01	1.07	1.16	1.25	1.39	1.56	1.72
2.5	1.02	1.04	1.15	1.25	1.34	1.47	1.63	1.77
2.00	1.06	1.12	1.26	1.36	1.45	1.56	1.70	1.81
1.90	1.08	1.15	1.28	1.39	1.47	1.58	1.71	1.82
1.80	1.10	1.17	1.31	1.41	1.49	1.60	1.72	1.83
1.70	1.12	1.20	1.34	1.44	1.52	1.62	1.74	1.84
1.60	1.15	1.23	1.37	1.47	1.54	1.64	1.75	1.85
1.50	1.18	1.26	1.40	1.49	1.57	1.66	1.77	1.86
1.40	1.21	1.30	1.43	1.52	1.59	1.68	1.78	1.87
1.30	1.25	1.33	1.46	1.55	1.62	1.70	1.80	1.87
1.20	1.29	1.37	1.50	1.58	1.64	1.72	1.81	1.88
1.10	1.33	1.41	1.53	1.61	1.67	1.75	1.83	1.89
1.00	1.38	1.46	1.57	1.64	1.70	1.77	1.84	1.90
0.90	1.43	1.50	1.61	1.68	1.73	1.79	1.86	1.91
0.80	1.48	1.55	1.65	1.71	1.75	1.81	1.87	1.92
0.70	1.53	1.60	1.69	1.74	1.78	1.83	1.89	1.93
0.60	1.59	1.65	1.73	1.78	1.81	1.86	1.90	1.94
0.50	1.65	1.70	1.77	1.81	1.84	1.88	1.92	1.95
0.40	1.71	1.75	1.81	1.84	1.87	1.90	1.93	1.96
0.30	1.76	1.80	1.85	1.87	1.89	1.92	1.95	1.97
0.20	1.82	1.84	1.88	1.90	1.92	1.94	1.96	1.98
0.10	1.86	1.88	1.91	1.93	1.94	1.96	1.97	1.98
0.00	1.90	1.92	1.94	1.95	1.96	1.97	1.98	1.99
–0.10	1.94	1.95	1.96	1.97	1.97	1.98	1.99	1.99
–0.20	1.96	1.97	1.97	1.98	1.98	1.99	1.99	2.00
–0.30	1.98	1.98	1.98	1.99	1.99	1.99	2.00	2.00
–0.40	1.99	1.99	1.99	1.99	1.99	2.00	2.00	2.00
–0.50	1.99	1.99	2.00	2.00	2.00	2.00	2.00	2.00

Table 5: The Effective Cost of Production when the cost of rework is 80 percent of the cost of scrap and the average is on the SCRAP side of the target

	C_p values									
C_{pk}	***0.10***	***0.20***	***0.30***	***0.40***	***0.50***	***0.60***	***0.70***	***0.80***	***0.90***	***1.00***
–0.50	15.44	15.38	15.43	15.49	15.56	15.61	15.65	15.68	15.71	15.73
–0.40	9.17	9.09	9.12	9.18	9.24	9.29	9.33	9.37	9.40	9.42
–0.30	5.93	5.82	5.83	5.88	5.94	5.99	6.04	6.07	6.11	6.13
–0.20	4.16	4.04	4.02	4.06	4.11	4.16	4.21	4.25	4.28	4.31
–0.10	3.16	3.02	2.97	2.99	3.04	3.09	3.13	3.17	3.21	3.24
0.00	2.58	2.42	2.35	2.34	2.38	2.42	2.47	2.51	2.55	2.58
0.10	2.23	2.06	1.97	1.94	1.95	1.99	2.03	2.08	2.11	2.15
0.20	-	1.86	1.74	1.68	1.68	1.70	1.74	1.78	1.82	1.85
0.30	-	-	1.61	1.53	1.50	1.50	1.53	1.57	1.61	1.64
0.40	-	-	-	1.45	1.39	1.37	1.39	1.41	1.45	1.48
0.50	-	-	-	-	1.33	1.29	1.29	1.30	1.33	1.36
0.60	-	-	-	-	-	1.25	1.22	1.22	1.24	1.27
0.70	-	-	-	-	-	-	1.19	1.17	1.18	1.20
0.80	-	-	-	-	-	-	-	1.15	1.14	1.15
0.90	-	-	-	-	-	-	-	-	1.12	1.11
1.00	-	-	-	-	-	-	-	-	-	1.10

Blank spaces correspond to impossible combinations of C_p and C_{pk}.

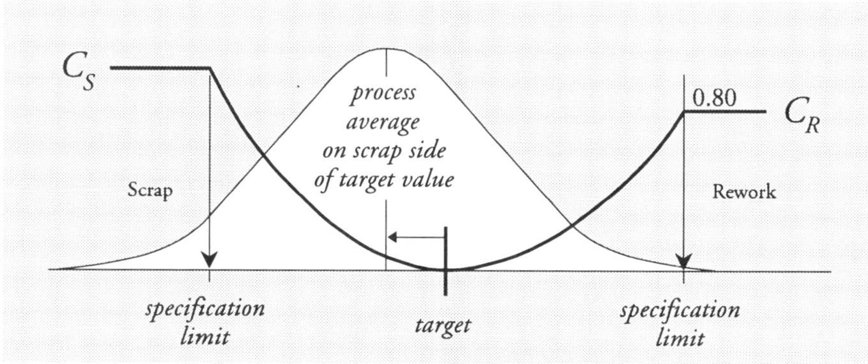

Table 5: Effective Cost of Production
When Cost of Rework is 80% the Cost of Scrap
and the Average is on the Scrap Side of the Target Value

Table 5: The Effective Cost of Production when the cost of rework is 80 percent of the cost of scrap and the average is on the SCRAP side of the target

C_{pk}	C_p values 1.10	1.20	1.30	1.40	1.50	1.60	1.70	1.80	1.90	2.00
–0.50	15.75	15.77	15.78	15.80	15.81	15.82	15.83	15.83	15.84	15.85
–0.40	9.44	9.46	9.48	9.49	9.51	9.52	9.53	9.53	9.54	9.55
–0.30	6.16	6.18	6.19	6.21	6.22	6.23	6.24	6.25	6.26	6.27
–0.20	4.33	4.36	4.38	4.39	4.41	4.42	4.43	4.44	4.45	4.46
–0.10	3.27	3.29	3.31	3.33	3.35	3.36	3.37	3.39	3.40	3.41
0.00	2.61	2.63	2.66	2.68	2.70	2.71	2.73	2.74	2.75	2.76
0.10	2.18	2.21	2.23	2.25	2.27	2.29	2.31	2.32	2.34	2.35
0.20	1.89	1.92	1.94	1.97	1.99	2.01	2.03	2.04	2.06	2.07
0.30	1.68	1.71	1.74	1.76	1.78	1.81	1.83	1.84	1.86	1.87
0.40	1.52	1.55	1.58	1.61	1.63	1.66	1.68	1.70	1.71	1.73
0.50	1.40	1.43	1.46	1.49	1.51	1.54	1.56	1.58	1.60	1.62
0.60	1.30	1.33	1.36	1.39	1.42	1.44	1.47	1.49	1.51	1.53
0.70	1.22	1.25	1.28	1.31	1.34	1.36	1.39	1.41	1.43	1.46
0.80	1.16	1.19	1.21	1.24	1.27	1.29	1.32	1.34	1.37	1.39
0.90	1.12	1.14	1.16	1.18	1.21	1.24	1.26	1.28	1.31	1.33
1.00	1.09	1.10	1.12	1.14	1.16	1.18	1.21	1.23	1.26	1.28
1.10	1.08	1.08	1.09	1.10	1.12	1.14	1.16	1.19	1.21	1.23
1.20	-	1.07	1.07	1.08	1.09	1.11	1.12	1.15	1.17	1.19
1.30	-	-	1.06	1.06	1.07	1.08	1.09	1.11	1.13	1.15
1.40	-	-	-	1.05	1.05	1.06	1.07	1.08	1.10	1.12
1.50	-	-	-	-	1.04	1.04	1.05	1.06	1.07	1.09
1.60	-	-	-	-	-	1.04	1.04	1.05	1.06	1.07
1.70	-	-	-	-	-	-	1.03	1.04	1.04	1.05
1.80	-	-	-	-	-	-	-	1.03	1.03	1.04
1.90	-	-	-	-	-	-	-	-	1.03	1.03
2.00	-	-	-	-	-	-	-	-	-	1.02

When $C_{pk} = C_p$ the average is at the target.

Table 5: The Effective Cost of Production when the cost of rework is 80 percent of the cost of scrap and the average is on the SCRAP side of the target

C_{pk}	C_p values 2.5	3.0	4.0	5.0	6.0	8.0	12.0	20.0
–0.50	15.91	15.93	15.95	15.97	15.98	15.99	16.00	16.01
–0.40	9.59	9.61	9.64	9.65	9.66	9.68	9.69	9.70
–0.30	6.31	6.33	6.36	6.38	6.39	6.40	6.42	6.43
–0.20	4.50	4.52	4.56	4.57	4.59	4.60	4.62	4.63
–0.10	3.45	3.48	3.51	3.53	3.55	3.56	3.58	3.60
0.00	2.81	2.84	2.88	2.90	2.92	2.94	2.96	2.98
0.10	2.40	2.43	2.48	2.50	2.52	2.55	2.57	2.59
0.20	2.12	2.16	2.21	2.24	2.27	2.29	2.32	2.34
0.30	1.94	1.98	2.04	2.07	2.10	2.13	2.16	2.19
0.40	1.80	1.85	1.91	1.95	1.98	2.02	2.05	2.08
0.50	1.70	1.75	1.82	1.87	1.90	1.94	1.98	2.02
0.60	1.61	1.67	1.75	1.81	1.84	1.89	1.94	1.98
0.70	1.54	1.61	1.70	1.76	1.80	1.85	1.90	1.95
0.80	1.48	1.55	1.65	1.72	1.76	1.82	1.88	1.93
0.90	1.43	1.50	1.61	1.68	1.73	1.79	1.86	1.92
1.00	1.38	1.46	1.57	1.65	1.70	1.77	1.84	1.90
1.10	1.33	1.41	1.53	1.61	1.67	1.75	1.83	1.89
1.20	1.29	1.37	1.50	1.58	1.64	1.72	1.81	1.88
1.30	1.25	1.33	1.46	1.55	1.62	1.70	1.80	1.87
1.40	1.21	1.30	1.43	1.52	1.59	1.68	1.78	1.87
1.50	1.18	1.26	1.40	1.49	1.57	1.66	1.77	1.86
1.60	1.15	1.23	1.37	1.47	1.54	1.64	1.75	1.85
1.70	1.12	1.20	1.34	1.44	1.52	1.62	1.74	1.84
1.80	1.10	1.17	1.31	1.41	1.49	1.60	1.72	1.83
1.90	1.08	1.15	1.28	1.39	1.47	1.58	1.71	1.82
2.00	1.06	1.12	1.26	1.36	1.45	1.56	1.70	1.81
2.5	1.02	1.04	1.15	1.25	1.34	1.47	1.63	1.77
3.0	-	1.01	1.07	1.16	1.25	1.39	1.56	1.72
4.0	-	-	1.01	1.04	1.11	1.25	1.45	1.64
5.0	-	-	-	1.00	1.03	1.14	1.34	1.56
6.0	-	-	-	-	1.00	1.06	1.25	1.49
8.0	-	-	-	-	-	1.00	1.11	1.36
12.0	-	-	-	-	-	-	1.00	1.16
20.0	-	-	-	-	-	-	-	1.00

Table 6: The Effective Cost of Production when the cost of rework is 80 percent of the cost of scrap and the average is on the REWORK side of the target

C_{pk}	C_p values 0.10	0.20	0.30	0.40	0.50	0.60	0.70	0.80	0.90	1.00
1.00	-	-	-	-	-	-	-	-	-	1.10
0.90	-	-	-	-	-	-	-	-	1.12	1.10
0.80	-	-	-	-	-	-	-	1.15	1.13	1.12
0.70	-	-	-	-	-	-	1.19	1.16	1.15	1.16
0.60	-	-	-	-	-	1.25	1.20	1.18	1.19	1.21
0.50	-	-	-	-	1.33	1.25	1.23	1.23	1.25	1.27
0.40	-	-	-	1.45	1.33	1.28	1.28	1.30	1.32	1.34
0.30	-	-	1.61	1.43	1.35	1.34	1.35	1.37	1.40	1.42
0.20	-	1.86	1.56	1.44	1.41	1.41	1.43	1.45	1.47	1.50
0.10	2.23	1.75	1.56	1.49	1.48	1.49	1.51	1.53	1.55	1.57
0.00	2.02	1.71	1.58	1.55	1.55	1.57	1.59	1.60	1.62	1.63
–0.10	1.90	1.70	1.63	1.62	1.62	1.64	1.65	1.66	1.67	1.68
–0.20	1.84	1.71	1.67	1.67	1.68	1.69	1.70	1.71	1.72	1.73
–0.30	1.81	1.73	1.72	1.72	1.73	1.73	1.74	1.75	1.75	1.75
–0.40	1.80	1.75	1.75	1.75	1.76	1.76	1.77	1.77	1.77	1.77
–0.50	1.79	1.77	1.77	1.77	1.78	1.78	1.78	1.78	1.79	1.79

Blank spaces correspond to impossible combinations of C_p and C_{pk}.

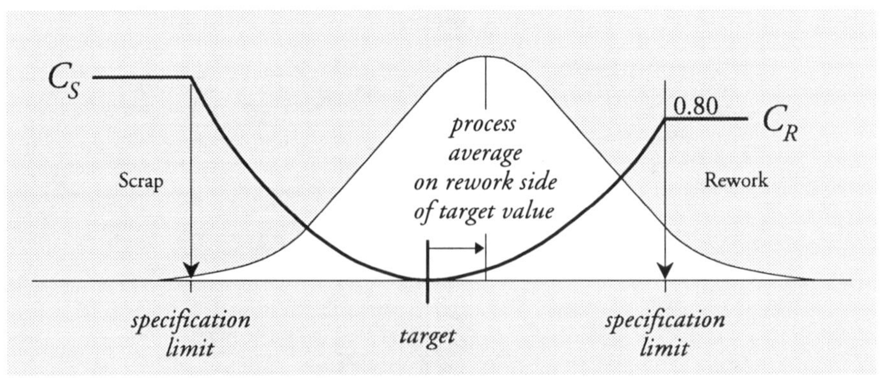

Table 6: Effective Cost of Production
When Cost of Rework is 80% the Cost of Scrap
and the Average is on the Rework Side of the Target Value

Table 6: The Effective Cost of Production when the cost of rework is 80 percent of the cost of scrap and the average is on the REWORK side of the target

C_{pk}	C_p values 1.10	1.20	1.30	1.40	1.50	1.60	1.70	1.80	1.90	2.00
2.00	-	-	-	-	-	-	-	-	-	1.02
1.90	-	-	-	-	-	-	-	-	1.03	1.03
1.80	-	-	-	-	-	-	-	1.03	1.03	1.03
1.70	-	-	-	-	-	-	1.03	1.03	1.03	1.04
1.60	-	-	-	-	-	1.04	1.04	1.04	1.05	1.05
1.50	-	-	-	-	1.04	1.04	1.04	1.05	1.06	1.07
1.40	-	-	-	1.05	1.05	1.05	1.06	1.07	1.08	1.09
1.30	-	-	1.06	1.05	1.06	1.06	1.08	1.09	1.10	1.12
1.20	-	1.07	1.06	1.06	1.07	1.09	1.10	1.12	1.13	1.15
1.10	1.08	1.07	1.07	1.08	1.10	1.11	1.13	1.15	1.17	1.18
1.00	1.09	1.09	1.10	1.11	1.13	1.15	1.17	1.19	1.20	1.22
0.90	1.10	1.11	1.13	1.15	1.17	1.19	1.21	1.23	1.25	1.26
0.80	1.13	1.15	1.17	1.19	1.21	1.23	1.25	1.27	1.29	1.31
0.70	1.18	1.20	1.22	1.24	1.26	1.29	1.331	1.32	1.34	1.36
0.60	1.23	1.25	1.28	1.30	1.32	1.34	1.36	1.38	1.39	1.41
0.50	1.30	1.32	1.34	1.36	1.38	1.40	1.42	1.44	1.45	1.46
0.40	1.37	1.39	1.41	1.43	1.45	1.46	1.48	1.49	1.51	1.52
0.30	1.44	1.46	1.48	1.50	1.51	1.53	1.54	1.55	1.56	1.57
0.20	1.51	1.53	1.55	1.56	1.57	1.58	1.59	1.60	1.61	1.62
0.10	1.58	1.60	1.61	1.62	1.63	1.64	1.65	1.65	1.66	1.67
0.00	1.64	1.65	1.66	1.67	1.68	1.68	1.69	1.69	1.70	1.70
–0.10	1.69	1.70	1.71	1.71	1.72	1.72	1.73	1.73	1.73	1.73
–0.20	1.73	1.74	1.74	1.74	1.75	1.75	1.75	1.75	1.76	1.76
–0.30	1.76	1.76	1.76	1.77	1.77	1.77	1.77	1.77	1.77	1.77
–0.40	1.78	1.78	1.78	1.78	1.78	1.78	1.78	1.78	1.79	1.79
–0.50	1.79	1.79	1.79	1.79	1.79	1.79	1.79	1.79	1.79	1.79

When $C_{pk} = C_p$ the average is at the target.

Table 6: The Effective Cost of Production when the cost of rework is 80 percent of the cost of scrap and the average is on the REWORK side of the target

	C_p values							
C_{pk}	**2.5**	**3.0**	**4.0**	**5.0**	**6.0**	**8.0**	**12.0**	**20.0**
20.0	-	-	-	-	-	-	-	1.00
12.0	-	-	-	-	-	-	1.00	1.13
8.0	-	-	-	-	-	1.00	1.09	1.29
6.0	-	-	-	-	1.00	1.05	1.20	1.39
5.0	-	-	-	1.00	1.02	1.11	1.27	1.45
4.0	-	-	1.01	1.04	1.09	1.20	1.36	1.51
3.0	-	1.01	1.06	1.13	1.20	1.31	1.45	1.58
2.5	1.02	1.03	1.12	1.20	1.27	1.38	1.50	1.61
2.00	1.05	1.10	1.21	1.29	1.36	1.45	1.56	1.65
1.90	1.06	1.12	1.23	1.31	1.38	1.47	1.57	1.66
1.80	1.08	1.14	1.25	1.33	1.39	1.48	1.58	1.66
1.70	1.10	1.16	1.27	1.35	1.41	1.50	1.59	1.67
1.60	1.12	1.18	1.29	1.37	1.43	1.51	1.60	1.68
1.50	1.14	1.21	1.32	1.40	1.45	1.53	1.61	1.68
1.40	1.17	1.24	1.34	1.42	1.47	1.55	1.62	1.69
1.30	1.12	1.27	1.37	1.44	1.49	1.56	1.64	1.70
1.20	1.23	1.30	1.40	1.47	1.51	1.58	1.65	1.71
1.10	1.27	1.33	1.43	1.49	1.54	1.60	1.66	1.71
1.00	1.30	1.37	1.46	1.52	1.56	1.61	1.67	1.72
0.90	1.34	1.40	1.49	1.54	1.58	1.63	1.69	1.73
0.80	1.38	1.44	1.52	1.57	1.60	1.65	1.70	1.74
0.70	1.43	1.48	1.55	1.59	1.63	1.67	1.71	1.75
0.60	1.47	1.52	1.58	1.62	1.65	1.69	1.72	1.75
0.50	1.52	1.56	1.61	1.65	1.67	1.70	1.73	1.76
0.40	1.57	1.60	1.65	1.67	1.69	1.72	1.75	1.77
0.30	1.61	1.64	1.68	1.70	1.72	1.74	1.76	1.77
0.20	1.65	1.68	1.70	1.72	1.74	1.75	1.77	1.78
0.10	1.69	1.71	1.73	1.74	1.75	1.76	1.78	1.79
0.00	1.72	1.73	1.75	1.76	1.77	1.77	1.78	1.79
–0.10	1.75	1.76	1.77	1.77	1.78	1.78	1.79	1.79
–0.20	1.77	1.77	1.78	1.78	1.79	1.79	1.79	1.80
–0.30	1.78	1.78	1.79	1.79	1.79	1.79	1.80	1.80
–0.40	1.79	1.79	1.79	1.79	1.80	1.80	1.80	1.80
–0.50	1.79	1.80	1.80	1.80	1.80	1.80	1.80	1.80

Table 7: The Effective Cost of Production when the cost of rework is 67 percent of the cost of scrap and the average is on the SCRAP side of the target

C_{pk}	C_p values 0.10	0.20	0.30	0.40	0.50	0.60	0.70	0.80	0.90	1.00
–0.50	15.39	15.37	15.42	15.49	15.56	15.61	15.65	15.68	15.71	15.73
–0.40	9.12	9.07	9.11	9.18	9.24	9.29	9.33	9.37	9.40	9.42
–0.30	5.87	5.80	5.82	5.88	5.94	5.99	6.04	6.07	6.11	6.13
–0.20	4.10	4.01	4.01	4.05	4.11	4.16	4.21	4.25	4.28	4.31
–0.10	3.09	2.98	2.96	2.99	3.04	3.09	3.13	3.17	3.21	3.24
0.00	2.49	2.37	2.33	2.34	2.37	2.42	2.47	2.51	2.55	2.58
0.10	2.14	2.01	1.94	1.93	1.95	1.99	2.03	2.08	2.11	2.15
0.20	-	1.80	1.70	1.67	1.67	1.70	1.74	1.78	1.82	1.85
0.30	-	-	1.57	1.50	1.49	1.50	1.53	1.57	1.61	1.64
0.40	-	-	-	1.42	1.37	1.36	1.38	1.41	1.45	1.48
0.50	-	-	-	-	1.31	1.28	1.28	1.30	1.33	1.36
0.60	-	-	-	-	-	1.23	1.21	1.22	1.24	1.27
0.70	-	-	-	-	-	-	1.18	1.17	1.18	1.20
0.80	-	-	-	-	-	-	-	1.14	1.13	1.14
0.90	-	-	-	-	-	-	-	-	1.11	1.11
1.00	-	-	-	-	-	-	-	-	-	1.09

Blank spaces correspond to impossible combinations of C_p and C_{pk}.

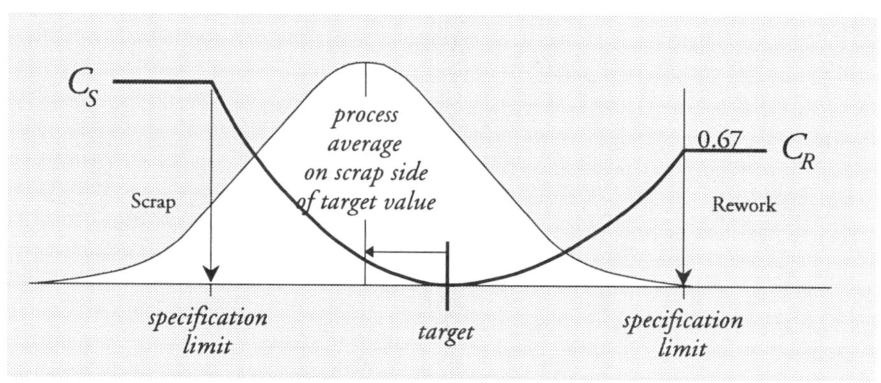

Table 7: Effective Cost of Production
When Cost of Rework is 67% the Cost of Scrap
and the Average is on the Scrap Side of the Target Value

Table 7: The Effective Cost of Production when the cost of rework is 67 percent of the cost of scrap and the average is on the SCRAP side of the target

C_{pk}	C_p values 1.10	1.20	1.30	1.40	1.50	1.60	1.70	1.80	1.90	2.00
–0.50	15.75	15.77	15.78	15.80	15.81	15.82	15.83	15.83	15.84	15.85
–0.40	9.44	9.46	9.48	9.49	9.51	9.52	9.53	9.53	9.54	9.55
–0.30	6.16	6.18	6.19	6.21	6.22	6.23	6.24	6.25	6.26	6.27
–0.20	4.33	4.36	4.38	4.39	4.41	4.42	4.43	4.44	4.45	4.46
–0.10	3.27	3.29	3.31	3.33	3.35	3.36	3.37	3.39	3.40	3.41
0.00	2.61	2.63	2.66	2.68	2.70	2.71	2.73	2.74	2.75	2.76
0.10	2.18	2.21	2.23	2.25	2.27	2.29	2.31	2.32	2.34	2.35
0.20	1.89	1.92	1.94	1.97	1.99	2.01	2.03	2.04	2.06	2.07
0.30	1.68	1.71	1.74	1.76	1.78	1.81	1.83	1.84	1.86	1.87
0.40	1.52	1.55	1.58	1.61	1.63	1.66	1.68	1.70	1.71	1.73
0.50	1.40	1.43	1.46	1.49	1.51	1.54	1.56	1.58	1.60	1.62
0.60	1.30	1.33	1.36	1.39	1.42	1.44	1.47	1.49	1.51	1.53
0.70	1.22	1.25	1.28	1.31	1.34	1.36	1.39	1.41	1.43	1.46
0.80	1.16	1.19	1.21	1.24	1.27	1.29	1.32	1.34	1.37	1.39
0.90	1.12	1.14	1.16	1.18	1.21	1.24	1.26	1.28	1.31	1.33
1.00	1.09	1.10	1.12	1.14	1.16	1.18	1.21	1.23	1.26	1.28
1.10	1.08	1.08	1.09	1.10	1.12	1.14	1.16	1.19	1.21	1.23
1.20	-	1.06	1.07	1.07	1.09	1.11	1.12	1.15	1.17	1.19
1.30	-	-	1.05	1.06	1.06	1.08	1.09	1.11	1.13	1.15
1.40	-	-	-	1.05	1.05	1.06	1.07	1.08	1.10	1.12
1.50	-	-	-	-	1.04	1.04	1.05	1.06	1.07	1.09
1.60	-	-	-	-	-	1.04	1.04	1.04	1.05	1.07
1.70	-	-	-	-	-	-	1.03	1.03	1.04	1.05
1.80	-	-	-	-	-	-	-	1.03	1.03	1.04
1.90	-	-	-	-	-	-	-	-	1.03	1.03
2.00	-	-	-	-	-	-	-	-	-	1.02

When $C_{pk} = C_p$ the average is at the target.

Table 7: The Effective Cost of Production when the cost of rework is 67 percent of the cost of scrap and the average is on the SCRAP side of the target

C_{pk}	C_p values 2.5	3.0	4.0	5.0	6.0	8.0	12.0	20.0
–0.50	15.91	15.93	15.95	15.97	15.98	15.99	16.00	16.01
–0.40	9.59	9.61	9.64	9.65	9.66	9.68	9.69	9.70
–0.30	6.31	6.33	6.36	6.38	6.39	6.40	6.42	6.43
–0.20	4.50	4.52	4.56	4.57	4.59	4.60	4.62	4.63
–0.10	3.45	3.48	3.51	3.53	3.55	3.56	3.58	3.60
0.00	2.81	2.84	2.88	2.90	2.92	2.94	2.96	2.98
0.10	2.40	2.43	2.48	2.50	2.52	2.55	2.57	2.59
0.20	2.12	2.16	2.21	2.24	2.27	2.29	2.32	2.34
0.30	1.94	1.98	2.04	2.07	2.10	2.13	2.16	2.19
0.40	1.80	1.85	1.91	1.95	1.98	2.02	2.05	2.08
0.50	1.70	1.75	1.82	1.87	1.90	1.94	1.98	2.02
0.60	1.61	1.67	1.75	1.81	1.84	1.89	1.94	1.98
0.70	1.54	1.61	1.70	1.76	1.80	1.85	1.90	1.95
0.80	1.48	1.55	1.65	1.72	1.76	1.82	1.88	1.93
0.90	1.43	1.50	1.61	1.68	1.73	1.79	1.86	1.92
1.00	1.38	1.46	1.57	1.65	1.70	1.77	1.84	1.90
1.10	1.33	1.41	1.53	1.61	1.67	1.75	1.83	1.89
1.20	1.29	1.37	1.50	1.58	1.64	1.72	1.81	1.88
1.30	1.25	1.33	1.46	1.55	1.62	1.70	1.80	1.87
1.40	1.21	1.30	1.43	1.52	1.59	1.68	1.78	1.87
1.50	1.18	1.26	1.40	1.49	1.57	1.66	1.77	1.86
1.60	1.15	1.23	1.37	1.47	1.54	1.64	1.75	1.85
1.70	1.12	1.20	1.34	1.44	1.52	1.62	1.74	1.84
1.80	1.10	1.17	1.31	1.41	1.49	1.60	1.72	1.83
1.90	1.08	1.15	1.28	1.39	1.47	1.58	1.71	1.82
2.00	1.06	1.12	1.26	1.36	1.45	1.56	1.70	1.81
2.5	1.01	1.04	1.15	1.25	1.34	1.47	1.63	1.77
3.0	-	1.01	1.07	1.16	1.25	1.39	1.56	1.72
4.0	-	-	1.01	1.04	1.11	1.25	1.45	1.64
5.0	-	-	-	1.00	1.03	1.14	1.34	1.56
6.0	-	-	-	-	1.00	1.06	1.25	1.49
8.0	-	-	-	-	-	1.00	1.11	1.36
12.0	-	-	-	-	-	-	1.00	1.16
20.0	-	-	-	-	-	-	-	1.00

Table 8: The Effective Cost of Production when the cost of rework is 67 percent of the cost of scrap and the average is on the REWORK side of the target

C_{pk}	C_p values: 0.10	0.20	0.30	0.40	0.50	0.60	0.70	0.80	0.90	1.00
1.00	-	-	-	-	-	-	-	-	-	1.09
0.90	-	-	-	-	-	-	-	-	1.11	1.09
0.80	-	-	-	-	-	-	-	1.14	1.11	1.11
0.70	-	-	-	-	-	-	1.18	1.14	1.13	1.13
0.60	-	-	-	-	-	1.23	1.18	1.16	1.16	1.18
0.50	-	-	-	-	1.31	1.22	1.20	1.20	1.21	1.23
0.40	-	-	-	1.42	1.29	1.24	1.24	1.25	1.27	1.29
0.30	-	-	1.57	1.38	1.31	1.29	1.29	1.31	1.33	1.35
0.20	-	1.80	1.50	1.39	1.35	1.35	1.36	1.38	1.40	1.41
0.10	2.14	1.68	1.49	1.42	1.40	1.41	1.43	1.44	1.46	1.47
0.00	1.92	1.62	1.50	1.46	1.46	1.48	1.49	1.50	1.52	1.53
–0.10	1.80	1.60	1.53	1.52	1.52	1.53	1.54	1.55	1.56	1.57
–0.20	1.73	1.60	1.57	1.56	1.57	1.58	1.59	1.59	1.60	1.60
–0.30	1.69	1.62	1.60	1.60	1.61	1.61	1.62	1.62	1.63	1.63
–0.40	1.67	1.63	1.62	1.63	1.63	1.63	1.64	1.64	1.64	1.65
–0.50	1.66	1.64	1.64	1.64	1.65	1.65	1.65	1.65	1.65	1.66

Blank spaces correspond to impossible combinations of C_p and C_{pk}.

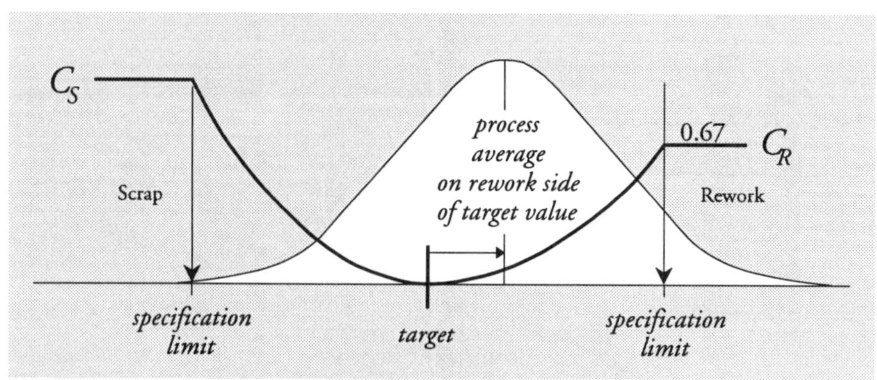

Table 8: Effective Cost of Production
When Cost of Rework is 67% the Cost of Scrap
and the Average is on the Rework Side of the Target Value

Table 8: The Effective Cost of Production when the cost of rework is 67 percent of the cost of scrap and the average is on the REWORK side of the target

C_{pk}	C_p values 1.10	1.20	1.30	1.40	1.50	1.60	1.70	1.80	1.90	2.00
2.00	-	-	-	-	-	-	-	-	-	1.02
1.90	-	-	-	-	-	-	-	-	1.03	1.02
1.80	-	-	-	-	-	-	-	1.03	1.03	1.03
1.70	-	-	-	-	-	-	1.03	1.03	1.03	1.03
1.60	-	-	-	-	-	1.04	1.03	1.03	1.04	1.05
1.50	-	-	-	-	1.04	1.04	1.04	1.04	1.05	1.06
1.40	-	-	-	1.05	1.04	1.04	1.05	1.06	1.07	1.08
1.30	-	-	1.05	1.05	1.05	1.05	1.06	1.07	1.09	1.10
1.20	-	1.06	1.05	1.05	1.06	1.07	1.08	1.10	1.11	1.13
1.10	1.08	1.06	1.06	1.07	1.08	1.09	1.11	1.12	1.14	1.15
1.00	1.08	1.07	1.08	1.09	1.11	1.12	1.14	1.15	1.17	1.19
0.90	1.09	1.10	1.11	1.12	1.14	1.16	1.17	1.19	1.20	1.22
0.80	1.11	1.13	1.14	1.16	1.18	1.19	1.21	1.23	1.24	1.26
0.70	1.15	1.17	1.18	1.20	1.22	1.24	1.25	1.27	1.28	1.30
0.60	1.19	1.21	1.23	1.25	1.27	1.29	1.30	1.32	1.33	1.34
0.50	1.25	1.27	1.29	1.30	1.32	1.34	1.35	1.36	1.38	1.39
0.40	1.31	1.32	1.34	1.36	1.37	1.39	1.40	1.41	1.42	1.43
0.30	1.37	1.38	1.40	1.41	1.43	1.44	1.45	1.46	1.47	1.48
0.20	1.43	1.44	1.46	1.47	1.48	1.49	1.50	1.50	1.51	1.52
0.10	1.49	1.50	1.51	1.52	1.52	1.53	1.54	1.54	1.55	1.55
0.00	1.54	1.54	1.55	1.56	1.56	1.57	1.57	1.58	1.58	1.59
–0.10	1.58	1.58	1.59	1.59	1.60	1.60	1.60	1.61	1.61	1.61
–0.20	1.61	1.61	1.62	1.62	1.62	1.62	1.63	1.63	1.63	1.63
–0.30	1.63	1.63	1.64	1.64	1.64	1.64	1.64	1.64	1.64	1.65
–0.40	1.65	1.65	1.65	1.65	1.65	1.65	1.65	1.65	1.65	1.65
–0.50	1.66	1.66	1.66	1.66	1.66	1.66	1.66	1.66	1.66	1.66

When $C_{pk} = C_p$ the average is at the target.

Table 8: The Effective Cost of Production when the cost of rework is 67 percent of the cost of scrap and the average is on the REWORK side of the target

	C_p values							
C_{pk}	***2.5***	***3.0***	***4.0***	***5.0***	***6.0***	***8.0***	***12.0***	***20.0***
20.0	-	-	-	-	-	-	-	1.00
12.0	-	-	-	-	-	-	1.00	1.11
8.0	-	-	-	-	-	1.00	1.07	1.24
6.0	-	-	-	-	1.00	1.04	1.17	1.33
5.0	-	-	-	1.00	1.02	1.09	1.23	1.38
4.0	-	-	1.01	1.03	1.08	1.17	1.30	1.43
3.0	-	1.01	1.05	1.11	1.17	1.26	1.38	1.48
2.5	1.01	1.03	1.10	1.17	1.23	1.32	1.42	1.51
2.00	1.04	1.08	1.17	1.24	1.30	1.38	1.46	1.54
1.90	1.05	1.10	1.19	1.26	1.31	1.39	1.47	1.55
1.80	1.06	1.11	1.21	1.28	1.33	1.40	1.48	1.55
1.70	1.08	1.13	1.23	1.29	1.34	1.41	1.49	1.56
1.60	1.10	1.15	1.24	1.31	1.36	1.43	1.50	1.56
1.50	1.12	1.17	1.27	1.33	1.38	1.44	1.51	1.57
1.40	1.14	1.20	1.29	1.35	1.39	1.45	1.52	1.58
1.30	1.17	1.22	1.31	1.37	1.41	1.47	1.53	1.58
1.20	1.19	1.25	1.33	1.39	1.43	1.48	1.54	1.59
1.10	1.22	1.28	1.36	1.41	1.45	1.50	1.55	1.60
1.00	1.25	1.30	1.38	1.43	1.47	1.51	1.56	1.60
0.90	1.28	1.33	1.40	1.45	1.48	1.53	1.57	1.61
0.80	1.32	1.37	1.43	1.47	1.50	1.54	1.58	1.61
0.70	1.36	1.40	1.46	1.50	1.52	1.56	1.59	1.62
0.60	1.39	1.43	1.48	1.52	1.54	1.57	1.60	1.63
0.50	1.43	1.47	1.51	1.54	1.56	1.59	1.61	1.63
0.40	1.47	1.50	1.54	1.56	1.58	1.60	1.62	1.64
0.30	1.51	1.53	1.56	1.58	1.60	1.61	1.63	1.65
0.20	1.54	1.56	1.59	1.60	1.61	1.63	1.64	1.65
0.10	1.58	1.59	1.61	1.62	1.63	1.64	1.65	1.65
0.00	1.60	1.61	1.63	1.63	1.64	1.65	1.65	1.66
–0.10	1.62	1.63	1.64	1.64	1.65	1.65	1.66	1.66
–0.20	1.64	1.64	1.65	1.65	1.66	1.66	1.66	1.66
–0.30	1.65	1.65	1.66	1.66	1.66	1.66	1.66	1.66
–0.40	1.66	1.66	1.66	1.66	1.66	1.66	1.66	1.67
–0.50	1.66	1.66	1.66	1.66	1.66	1.67	1.67	1.67

Table 9: The Effective Cost of Production when the cost of rework is 50 percent of the cost of scrap and the average is on the SCRAP side of the target

	C_p values									
C_{pk}	***0.10***	***0.20***	***0.30***	***0.40***	***0.50***	***0.60***	***0.70***	***0.80***	***0.90***	***1.00***
–0.50	15.34	15.35	15.42	15.49	15.56	15.61	15.65	15.68	15.71	15.73
–0.40	9.05	9.05	9.10	9.17	9.24	9.29	9.33	9.37	9.40	9.42
–0.30	5.80	5.77	5.81	5.88	5.94	5.99	6.04	6.07	6.11	6.13
–0.20	4.02	3.97	4.00	4.05	4.11	4.16	4.21	4.25	4.28	4.31
–0.10	3.00	2.94	2.94	2.98	3.03	3.09	3.13	3.17	3.21	3.24
0.00	2.39	2.32	2.30	2.33	2.37	2.42	2.47	2.51	2.55	2.58
0.10	2.02	1.94	1.91	1.91	1.94	1.99	2.03	2.08	2.11	2.15
0.20	-	1.72	1.66	1.64	1.66	1.70	1.74	1.78	1.82	1.85
0.30	-	-	1.51	1.47	1.47	1.49	1.53	1.57	1.61	1.64
0.40	-	-	-	1.37	1.35	1.36	1.38	1.41	1.45	1.48
0.50	-	-	-	-	1.28	1.26	1.27	1.30	1.33	1.36
0.60	-	-	-	-	-	1.21	1.20	1.21	1.24	1.27
0.70	-	-	-	-	-	-	1.16	1.16	1.17	1.19
0.80	-	-	-	-	-	-	-	1.13	1.13	1.14
0.90	-	-	-	-	-	-	-	-	1.10	1.10
1.00	-	-	-	-	-	-	-	-	-	1.08

Blank spaces correspond to impossible combinations of C_p and C_{pk}.

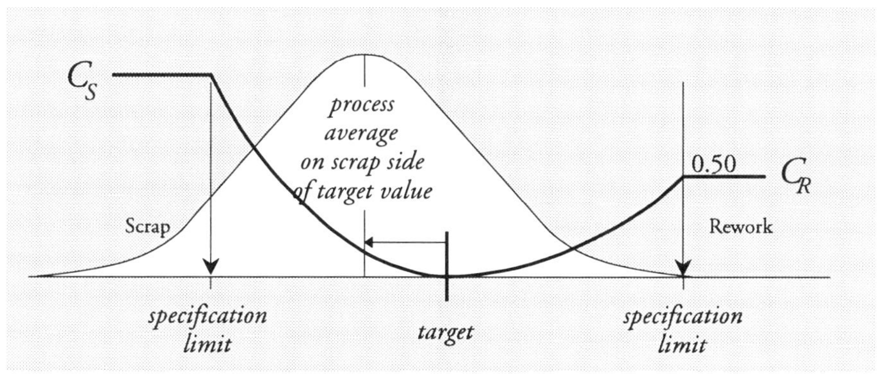

Table 9: Effective Cost of Production When Cost of Rework is 50% the Cost of Scrap and the Average is on the Scrap Side of the Target Value

Table 9: The Effective Cost of Production when the cost of rework is 50 percent of the cost of scrap and the average is on the SCRAP side of the target

C_{pk}	C_p values: 1.10	1.20	1.30	1.40	1.50	1.60	1.70	1.80	1.90	2.00
–0.50	15.75	15.77	15.78	15.80	15.81	15.82	15.83	15.83	15.84	15.85
–0.40	9.44	9.46	9.48	9.49	9.51	9.52	9.53	9.53	9.54	9.55
–0.30	6.16	6.18	6.19	6.21	6.22	6.23	6.24	6.25	6.26	6.27
–0.20	4.33	4.36	4.38	4.39	4.41	4.42	4.43	4.44	4.45	4.46
–0.10	3.27	3.29	3.31	3.33	3.35	3.36	3.37	3.39	3.40	3.41
0.00	2.61	2.63	2.66	2.68	2.70	2.71	2.73	2.74	2.75	2.76
0.10	2.18	2.21	2.23	2.25	2.27	2.29	2.31	2.32	2.34	2.35
0.20	1.89	1.92	1.94	1.97	1.99	2.01	2.03	2.04	2.06	2.07
0.30	1.68	1.71	1.74	1.76	1.78	1.81	1.83	1.84	1.86	1.87
0.40	1.52	1.55	1.58	1.61	1.63	1.66	1.68	1.70	1.71	1.73
0.50	1.40	1.43	1.46	1.49	1.51	1.54	1.56	1.58	1.60	1.62
0.60	1.30	1.33	1.36	1.39	1.42	1.44	1.47	1.49	1.51	1.53
0.70	1.22	1.25	1.28	1.31	1.34	1.36	1.39	1.41	1.43	1.46
0.80	1.16	1.19	1.21	1.24	1.27	1.29	1.32	1.34	1.37	1.39
0.90	1.12	1.14	1.16	1.18	1.21	1.24	1.26	1.28	1.31	1.33
1.00	1.09	1.10	1.12	1.14	1.16	1.18	1.21	1.23	1.26	1.28
1.10	1.07	1.07	1.08	1.10	1.12	1.14	1.16	1.19	1.21	1.23
1.20	-	1.06	1.06	1.07	1.09	1.10	1.12	1.15	1.17	1.19
1.30	-	-	1.05	1.05	1.06	1.08	1.09	1.11	1.13	1.15
1.40	-	-	-	1.04	1.05	1.06	1.07	1.08	1.10	1.12
1.50	-	-	-	-	1.04	1.04	1.05	1.06	1.07	1.09
1.60	-	-	-	-	-	1.03	1.04	1.04	1.05	1.07
1.70	-	-	-	-	-	-	1.03	1.03	1.04	1.05
1.80	-	-	-	-	-	-	-	1.03	1.03	1.04
1.90	-	-	-	-	-	-	-	-	1.02	1.03
2.00	-	-	-	-	-	-	-	-	-	1.02

When $C_{pk} = C_p$ the average is at the target.

Table 9: The Effective Cost of Production when the cost of rework is 50 percent of the cost of scrap and the average is on the SCRAP side of the target

	C_p values							
C_{pk}	**2.5**	**3.0**	**4.0**	**5.0**	**6.0**	**8.0**	**12.0**	**20.0**
–0.50	15.91	15.93	15.95	15.97	15.98	15.99	16.00	16.01
–0.40	9.59	9.61	9.64	9.65	9.66	9.68	9.69	9.70
–0.30	6.31	6.33	6.36	6.38	6.39	6.40	6.42	6.43
–0.20	4.50	4.52	4.56	4.57	4.59	4.60	4.62	4.63
–0.10	3.45	3.48	3.51	3.53	3.55	3.56	3.58	3.60
0.00	2.81	2.84	2.88	2.90	2.92	2.94	2.96	2.98
0.10	2.40	2.43	2.48	2.50	2.52	2.55	2.57	2.59
0.20	2.12	2.16	2.21	2.24	2.27	2.29	2.32	2.34
0.30	1.94	1.98	2.04	2.07	2.10	2.13	2.16	2.19
0.40	1.80	1.85	1.91	1.95	1.98	2.02	2.05	2.08
0.50	1.70	1.75	1.82	1.87	1.90	1.94	1.98	2.02
0.60	1.61	1.67	1.75	1.81	1.84	1.89	1.94	1.98
0.70	1.54	1.61	1.70	1.76	1.80	1.85	1.90	1.95
0.80	1.48	1.55	1.65	1.72	1.76	1.82	1.88	1.93
0.90	1.43	1.50	1.61	1.68	1.73	1.79	1.86	1.92
1.00	1.38	1.46	1.57	1.65	1.70	1.77	1.84	1.90
1.10	1.33	1.41	1.53	1.61	1.67	1.75	1.83	1.89
1.20	1.29	1.37	1.50	1.58	1.64	1.72	1.81	1.88
1.30	1.25	1.33	1.46	1.55	1.62	1.70	1.80	1.87
1.40	1.21	1.30	1.43	1.52	1.59	1.68	1.78	1.87
1.50	1.18	1.26	1.40	1.49	1.57	1.66	1.77	1.86
1.60	1.15	1.23	1.37	1.47	1.54	1.64	1.75	1.85
1.70	1.12	1.20	1.34	1.44	1.52	1.62	1.74	1.84
1.80	1.10	1.17	1.31	1.41	1.49	1.60	1.72	1.83
1.90	1.08	1.15	1.28	1.39	1.47	1.58	1.71	1.82
2.00	1.06	1.12	1.26	1.36	1.45	1.56	1.70	1.81
2.5	1.01	1.04	1.15	1.25	1.34	1.47	1.63	1.77
3.0	-	1.01	1.07	1.16	1.25	1.39	1.56	1.72
4.0	-	-	1.01	1.04	1.11	1.25	1.45	1.64
5.0	-	-	-	1.00	1.03	1.14	1.34	1.56
6.0	-	-	-	-	1.00	1.06	1.25	1.49
8.0	-	-	-	-	-	1.00	1.11	1.36
12.0	-	-	-	-	-	-	1.00	1.16
20.0	-	-	-	-	-	-	-	1.00

Table 10: The Effective Cost of Production when the cost of rework is 50 percent of the cost of scrap and the average is on the REWORK side of the target

C_{pk}	C_p values 0.10	0.20	0.30	0.40	0.50	0.60	0.70	0.80	0.90	1.00
1.00	-	-	-	-	-	-	-	-	-	1.08
0.90	-	-	-	-	-	-	-	-	1.10	1.08
0.80	-	-	-	-	-	-	-	1.13	1.09	1.08
0.70	-	-	-	-	-	-	1.16	1.12	1.10	1.10
0.60	-	-	-	-	-	1.21	1.15	1.13	1.12	1.13
0.50	-	-	-	-	1.28	1.19	1.16	1.15	1.16	1.17
0.40	-	-	-	1.37	1.25	1.20	1.18	1.19	1.20	1.21
0.30	-	-	1.51	1.32	1.25	1.22	1.22	1.23	1.25	1.26
0.20	-	1.72	1.43	1.31	1.27	1.26	1.27	1.28	1.30	1.31
0.10	2.02	1.58	1.40	1.33	1.31	1.31	1.32	1.33	1.34	1.35
0.00	1.80	1.51	1.39	1.36	1.35	1.36	1.37	1.38	1.39	1.39
–0.10	1.66	1.48	1.41	1.39	1.39	1.40	1.41	1.42	1.42	1.43
–0.20	1.58	1.47	1.43	1.42	1.43	1.43	1.44	1.44	1.45	1.45
–0.30	1.54	1.47	1.45	1.45	1.45	1.46	1.46	1.47	1.47	1.47
–0.40	1.51	1.48	1.47	1.47	1.47	1.48	1.48	1.48	1.48	1.48
–0.50	1.50	1.49	1.48	1.48	1.49	1.49	1.49	1.49	1.49	1.49

Blank spaces correspond to impossible combinations of C_p and C_{pk}.

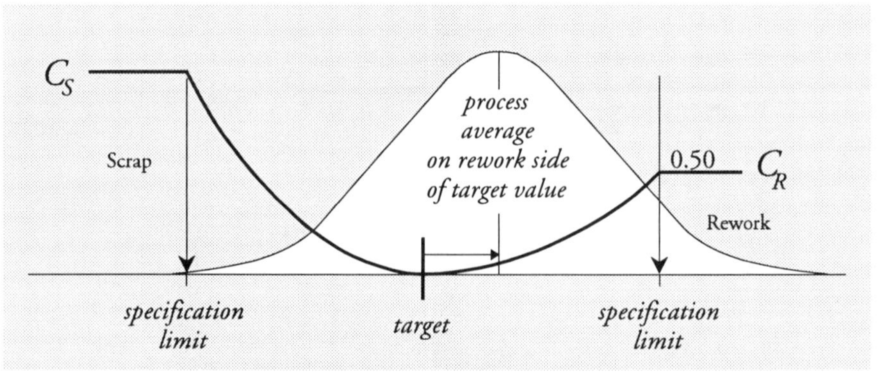

Table 10: Effective Cost of Production
When Cost of Rework is 50% the Cost of Scrap
and the Average is on the Rework Side of the Target Value

Table 10: The Effective Cost of Production when the cost of rework is 50 percent of the cost of scrap and the average is on the REWORK side of the target

	C_p values									
C_{pk}	**1.10**	**1.20**	**1.30**	**1.40**	**1.50**	**1.60**	**1.70**	**1.80**	**1.90**	**2.00**
2.00	-	-	-	-	-	-	-	-	-	1.02
1.90	-	-	-	-	-	-	-	-	1.02	1.02
1.80	-	-	-	-	-	-	-	1.03	1.02	1.02
1.70	-	-	-	-	-	-	1.03	1.02	1.02	1.03
1.60	-	-	-	-	-	1.03	1.03	1.03	1.03	1.03
1.50	-	-	-	-	1.04	1.04	1.03	1.03	1.04	1.05
1.40	-	-	-	1.04	1.03	1.03	1.04	1.04	1.05	1.06
1.30	-	-	1.05	1.04	1.04	1.04	1.05	1.06	1.07	1.08
1.20	-	1.06	1.05	1.04	1.05	1.05	1.06	1.07	1.08	1.09
1.10	1.07	1.05	1.05	1.05	1.06	1.07	1.08	1.09	1.10	1.12
1.00	1.06	1.06	1.06	1.07	1.08	1.09	1.10	1.12	1.13	1.14
0.90	1.07	1.07	1.08	1.09	1.10	1.12	1.13	1.14	1.15	1.16
0.80	1.09	1.10	1.11	1.12	1.13	1.15	1.16	1.17	1.18	1.19
0.70	1.11	1.12	1.14	1.15	1.17	1.18	1.19	1.20	1.21	1.22
0.60	1.15	1.16	1.17	1.19	1.20	1.21	1.23	1.24	1.25	1.26
0.50	1.19	1.20	1.21	1.23	1.24	1.25	1.26	1.27	1.28	1.29
0.40	1.23	1.24	1.26	1.27	1.28	1.29	1.30	1.31	1.32	1.32
0.30	1.28	1.29	1.30	1.31	1.32	1.33	1.34	1.34	1.35	1.36
0.20	1.32	1.33	1.34	1.35	1.36	1.37	1.37	1.38	1.38	1.39
0.10	1.36	1.37	1.38	1.39	1.39	1.40	1.40	1.41	1.41	1.42
0.00	1.40	1.41	1.41	1.42	1.42	1.43	1.43	1.43	1.44	1.44
–0.10	1.43	1.44	1.44	1.44	1.45	1.45	1.45	1.46	1.46	1.46
–0.20	1.46	1.46	1.46	1.46	1.47	1.47	1.47	1.47	1.47	1.47
–0.30	1.47	1.48	1.48	1.48	1.48	1.48	1.48	1.48	1.48	1.48
–0.40	1.49	1.49	1.49	1.49	1.49	1.49	1.49	1.49	1.49	1.49
–0.50	1.49	1.49	1.49	1.49	1.49	1.49	1.49	1.49	1.50	1.50

When $C_{pk} = C_p$ the average is at the target.

Table 10: The Effective Cost of Production when the cost of rework is 50 percent of the cost of scrap and the average is on the REWORK side of the target

C_{pk}	C_p values 2.5	3.0	4.0	5.0	6.0	8.0	12.0	20.0
20.0	-	-	-	-	-	-	-	1.00
12.0	-	-	-	-	-	-	1.00	1.08
8.0	-	-	-	-	-	1.00	1.06	1.18
6.0	-	-	-	-	1.00	1.03	1.13	1.25
5.0	-	-	-	1.00	1.02	1.07	1.17	1.28
4.0	-	-	1.01	1.02	1.06	1.13	1.22	1.32
3.0	-	1.01	1.03	1.28	1.13	1.20	1.28	1.36
2.5	1.01	1.02	1.07	1.13	1.17	1.24	1.31	1.38
2.00	1.03	1.06	1.13	1.18	1.22	1.28	1.35	1.41
1.90	1.04	1.07	1.14	1.19	1.24	1.29	1.35	1.41
1.80	1.05	1.09	1.15	1.21	1.25	1.30	1.36	1.41
1.70	1.06	1.10	1.17	1.22	1.26	1.31	1.37	1.42
1.60	1.07	1.12	1.18	1.23	1.27	1.32	1.38	1.42
1.50	1.09	1.13	1.20	1.25	1.28	1.33	1.38	1.43
1.40	1.11	1.15	1.21	1.26	1.30	1.34	1.39	1.43
1.30	1.12	1.17	1.23	1.28	1.31	1.35	1.40	1.44
1.20	1.14	1.19	1.25	1.29	1.32	1.36	1.41	1.44
1.10	1.17	1.21	1.27	1.31	1.34	1.37	1.41	1.45
1.00	1.19	1.23	1.28	1.32	1.35	1.38	1.42	1.45
0.90	1.21	1.25	1.30	1.34	1.36	1.39	1.43	1.46
0.80	1.24	1.27	1.32	1.35	1.38	1.41	1.44	1.46
0.70	1.27	1.30	1.34	1.37	1.39	1.42	1.44	1.47
0.60	1.30	1.32	1.36	1.39	1.41	1.43	1.45	1.47
0.50	1.32	1.35	1.38	1.41	1.42	1.44	1.46	1.48
0.40	1.35	1.38	1.40	1.42	1.43	1.45	1.47	1.48
0.30	1.38	1.40	1.42	1.44	1.45	1.46	1.47	1.48
0.20	1.41	1.42	1.44	1.45	1.46	1.47	1.48	1.49
0.10	1.43	1.44	1.46	1.46	1.47	1.48	1.49	1.49
0.00	1.45	1.46	1.47	1.48	1.48	1.48	1.49	1.49
–0.10	1.47	1.47	1.48	1.48	1.49	1.49	1.49	1.50
–0.20	1.48	1.48	1.49	1.49	1.49	1.49	1.50	1.50
–0.30	1.49	1.49	1.49	1.49	1.49	1.50	1.50	1.50
–0.40	1.49	1.49	1.50	1.50	1.50	1.50	1.50	1.50
–0.50	1.50	1.50	1.50	1.50	1.50	1.50	1.50	1.50

Table 11: The Effective Cost of Production when the cost of rework is 33 percent of the cost of scrap and the average is on the SCRAP side of the target

C_{pk}	C_p values 0.10	0.20	0.30	0.40	0.50	0.60	0.70	0.80	0.90	1.00
–0.50	15.28	15.33	15.41	15.49	15.56	15.61	15.65	15.68	15.71	15.73
–0.40	8.99	9.03	9.10	9.17	9.24	9.29	9.33	9.37	9.40	9.42
–0.30	5.72	5.75	5.80	5.87	5.94	5.99	6.04	6.07	6.11	6.13
–0.20	3.93	3.94	3.98	4.05	4.11	4.16	4.21	4.25	4.28	4.31
–0.10	2.90	2.89	2.92	2.98	3.03	3.09	3.13	3.17	3.21	3.24
0.00	2.29	2.27	2.28	2.32	2.37	2.42	2.47	2.51	2.55	2.58
0.10	1.91	1.88	1.87	1.90	1.94	1.99	2.03	2.08	2.11	2.15
0.20	-	1.64	1.62	1.62	1.65	1.69	1.74	1.78	1.82	1.85
0.30	-	-	1.46	1.45	1.46	1.49	1.53	1.57	1.60	1.64
0.40	-	-	-	1.33	1.33	1.35	1.38	1.41	1.45	1.48
0.50	-	-	-	-	1.25	1.25	1.27	1.29	1.33	1.36
0.60	-	-	-	-	-	1.19	1.19	1.21	1.24	1.27
0.70	-	-	-	-	-	-	1.14	1.15	1.17	1.19
0.80	-	-	-	-	-	-	-	1.11	1.12	1.14
0.90	-	-	-	-	-	-	-	-	1.09	1.10
1.00	-	-	-	-	-	-	-	-	-	1.07

Blank spaces correspond to impossible combinations of C_p and C_{pk}.

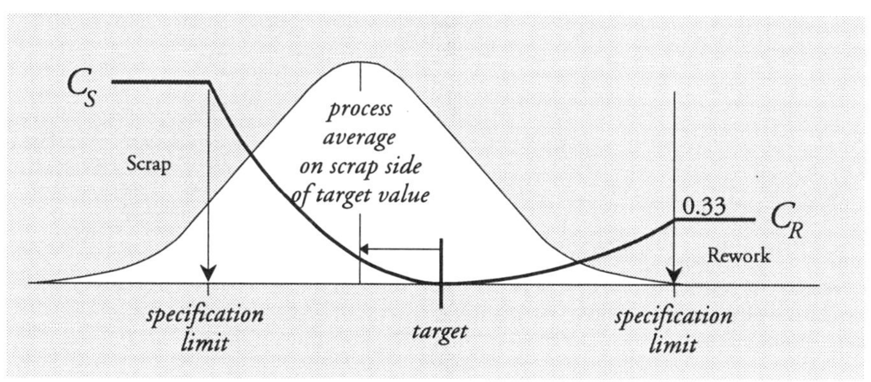

Table 11: Effective Cost of Production
When Cost of Rework is 33% the Cost of Scrap
and the Average is on the Scrap Side of the Target Value

Table 11: The Effective Cost of Production when the cost of rework is 33 percent of the cost of scrap and the average is on the SCRAP side of the target

C_{pk}	C_p values: 1.10	1.20	1.30	1.40	1.50	1.60	1.70	1.80	1.90	2.00
–0.50	15.75	15.77	15.78	15.80	15.81	15.82	15.83	15.83	15.84	15.85
–0.40	9.44	9.46	9.48	9.49	9.51	9.52	9.53	9.53	9.54	9.55
–0.30	6.16	6.18	6.19	6.21	6.22	6.23	6.24	6.25	6.26	6.27
–0.20	4.33	4.36	4.38	4.39	4.41	4.42	4.43	4.44	4.45	4.46
–0.10	3.27	3.29	3.31	3.33	3.35	3.36	3.37	3.39	3.40	3.41
0.00	2.61	2.63	2.66	2.68	2.70	2.71	2.73	2.74	2.75	2.76
0.10	2.18	2.21	2.23	2.25	2.27	2.29	2.31	2.32	2.34	2.35
0.20	1.89	1.92	1.94	1.97	1.99	2.01	2.03	2.04	2.06	2.07
0.30	1.68	1.71	1.74	1.76	1.78	1.81	1.83	1.84	1.86	1.87
0.40	1.52	1.55	1.58	1.61	1.63	1.66	1.68	1.70	1.71	1.73
0.50	1.40	1.43	1.46	1.49	1.51	1.54	1.56	1.58	1.60	1.62
0.60	1.30	1.33	1.36	1.39	1.42	1.44	1.47	1.49	1.51	1.53
0.70	1.22	1.25	1.28	1.31	1.34	1.36	1.39	1.41	1.43	1.46
0.80	1.16	1.19	1.21	1.24	1.27	1.29	1.32	1.34	1.37	1.39
0.90	1.11	1.13	1.16	1.18	1.21	1.24	1.26	1.28	1.31	1.33
1.00	1.08	1.10	1.11	1.14	1.16	1.18	1.21	1.23	1.26	1.28
1.10	1.06	1.07	1.08	1.10	1.12	1.14	1.16	1.19	1.21	1.23
1.20	-	1.05	1.06	1.07	1.09	1.10	1.12	1.15	1.17	1.19
1.30	-	-	1.04	1.05	1.06	1.08	1.09	1.11	1.13	1.15
1.40	-	-	-	1.04	1.04	1.05	1.07	1.08	1.10	1.12
1.50	-	-	-	-	1.03	1.04	1.05	1.06	1.07	1.09
1.60	-	-	-	-	-	1.03	1.03	1.04	1.05	1.07
1.70	-	-	-	-	-	-	1.03	1.03	1.04	1.05
1.80	-	-	-	-	-	-	-	1.02	1.03	1.03
1.90	-	-	-	-	-	-	-	-	1.02	1.02
2.00	-	-	-	-	-	-	-	-	-	1.02

When $C_{pk} = C_p$ the average is at the target.

Table 11: The Effective Cost of Production when the cost of rework is 33 percent of the cost of scrap and the average is on the SCRAP side of the target

	C_p values							
C_{pk}	2.5	3.0	4.0	5.0	6.0	8.0	12.0	20.0
–0.50	15.91	15.93	15.95	15.97	15.98	15.99	16.00	16.01
–0.40	9.59	9.61	9.64	9.65	9.66	9.68	9.69	9.70
–0.30	6.31	6.33	6.36	6.38	6.39	6.40	6.42	6.43
–0.20	4.50	4.52	4.56	4.57	4.59	4.60	4.62	4.63
–0.10	3.45	3.48	3.51	3.53	3.55	3.56	3.58	3.60
0.00	2.81	2.84	2.88	2.90	2.92	2.94	2.96	2.98
0.10	2.40	2.43	2.48	2.50	2.52	2.55	2.57	2.59
0.20	2.12	2.16	2.21	2.24	2.27	2.29	2.32	2.34
0.30	1.94	1.98	2.04	2.07	2.10	2.13	2.16	2.19
0.40	1.80	1.85	1.91	1.95	1.98	2.02	2.05	2.08
0.50	1.70	1.75	1.82	1.87	1.90	1.94	1.98	2.02
0.60	1.61	1.67	1.75	1.81	1.84	1.89	1.94	1.98
0.70	1.54	1.61	1.70	1.76	1.80	1.85	1.90	1.95
0.80	1.48	1.55	1.65	1.72	1.76	1.82	1.88	1.93
0.90	1.43	1.50	1.61	1.68	1.73	1.79	1.86	1.92
1.00	1.38	1.46	1.57	1.65	1.70	1.77	1.84	1.90
1.10	1.33	1.41	1.53	1.61	1.67	1.75	1.83	1.89
1.20	1.29	1.37	1.50	1.58	1.64	1.72	1.81	1.88
1.30	1.25	1.33	1.46	1.55	1.62	1.70	1.80	1.87
1.40	1.21	1.30	1.43	1.52	1.59	1.68	1.78	1.87
1.50	1.18	1.26	1.40	1.49	1.57	1.66	1.77	1.86
1.60	1.15	1.23	1.37	1.47	1.54	1.64	1.75	1.85
1.70	1.12	1.20	1.34	1.44	1.52	1.62	1.74	1.84
1.80	1.10	1.17	1.31	1.41	1.49	1.60	1.72	1.83
1.90	1.08	1.15	1.28	1.39	1.47	1.58	1.71	1.82
2.00	1.06	1.12	1.26	1.36	1.45	1.56	1.70	1.81
2.5	1.01	1.04	1.15	1.25	1.34	1.47	1.63	1.77
3.0	-	1.01	1.07	1.16	1.25	1.39	1.56	1.72
4.0	-	-	1.00	1.04	1.11	1.25	1.45	1.64
5.0	-	-	-	1.00	1.03	1.14	1.34	1.56
6.0	-	-	-	-	1.00	1.06	1.25	1.49
8.0	-	-	-	-	-	1.00	1.11	1.36
12.0	-	-	-	-	-	-	1.00	1.16
20.0	-	-	-	-	-	-	-	1.00

Table 12: The Effective Cost of Production when the cost of rework is 33 percent of the cost of scrap and the average is on the REWORK side of the target

C_{pk}	C_p values 0.10	0.20	0.30	0.40	0.50	0.60	0.70	0.80	0.90	1.00
1.00	-	-	-	-	-	-	-	-	-	1.07
0.90	-	-	-	-	-	-	-	-	1.09	1.06
0.80	-	-	-	-	-	-	-	1.11	1.08	1.06
0.70	-	-	-	-	-	-	1.14	1.10	1.08	1.07
0.60	-	-	-	-	-	1.19	1.12	1.09	1.09	1.09
0.50	-	-	-	-	1.25	1.16	1.12	1.11	1.11	1.11
0.40	-	-	-	1.33	1.20	1.15	1.13	1.13	1.13	1.14
0.30	-	-	1.46	1.27	1.19	1.16	1.15	1.16	1.17	1.17
0.20	-	1.64	1.36	1.24	1.19	1.18	1.18	1.19	1.20	1.21
0.10	1.91	1.49	1.31	1.23	1.21	1.21	1.21	1.22	1.23	1.24
0.00	1.67	1.40	1.29	1.25	1.24	1.24	1.25	1.25	1.26	1.26
–0.10	1.53	1.35	1.28	1.26	1.26	1.27	1.27	1.28	1.28	1.29
–0.20	1.44	1.33	1.29	1.28	1.29	1.29	1.29	1.30	1.30	1.30
–0.30	1.39	1.32	1.30	1.30	1.30	1.31	1.31	1.31	1.31	1.31
–0.40	1.36	1.32	1.31	1.31	1.32	1.32	1.32	1.32	1.32	1.32
–0.50	1.34	1.33	1.32	1.32	1.32	1.32	1.33	1.33	1.33	1.33

Blank spaces correspond to impossible combinations of C_p and C_{pk}.

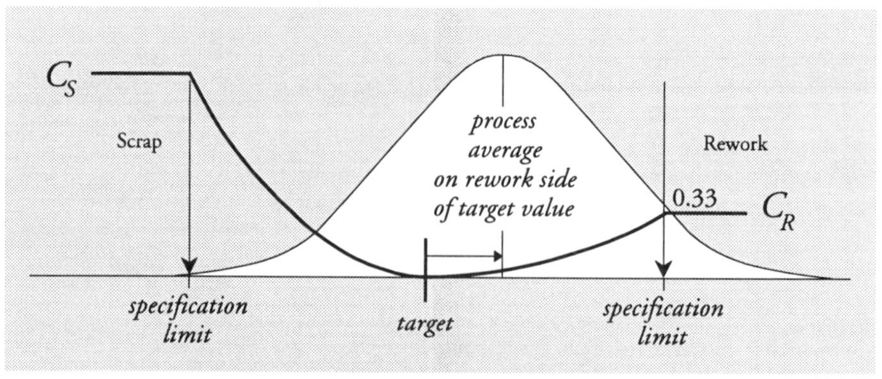

Table 12: Effective Cost of Production When Cost of Rework is 33% the Cost of Scrap and the Average is on the Rework Side of the Target Value

Table 12: The Effective Cost of Production when the cost of rework is 33 percent of the cost of scrap and the average is on the REWORK side of the target

C_{pk}	C_p values 1.10	1.20	1.30	1.40	1.50	1.60	1.70	1.80	1.90	2.00
2.00	-	-	-	-	-	-	-	-	-	1.02
1.90	-	-	-	-	-	-	-	-	1.02	1.02
1.80	-	-	-	-	-	-	-	1.02	1.02	1.02
1.70	-	-	-	-	-	-	1.03	1.02	1.02	1.02
1.60	-	-	-	-	-	1.03	1.02	1.02	1.02	1.02
1.50	-	-	-	-	1.03	1.02	1.02	1.02	1.03	1.03
1.40	-	-	-	1.04	1.03	1.02	1.03	1.03	1.03	1.04
1.30	-	-	1.04	1.03	1.03	1.03	1.03	1.04	1.04	1.05
1.20	-	1.05	1.04	1.03	1.03	1.04	1.04	1.05	1.06	1.06
1.10	1.06	1.04	1.04	1.04	1.04	1.05	1.05	1.06	1.07	1.08
1.00	1.05	1.04	1.04	1.05	1.05	1.06	1.07	1.08	1.09	1.09
0.90	1.05	1.05	1.06	1.06	1.07	1.08	1.09	1.09	1.10	1.11
0.80	1.06	1.06	1.07	1.08	1.09	1.10	1.11	1.11	1.12	1.13
0.70	1.08	1.08	1.09	1.10	1.11	1.12	1.13	1.14	1.14	1.15
0.60	1.10	1.11	1.12	1.13	1.13	1.14	1.15	1.16	1.16	1.17
0.50	1.12	1.13	1.14	1.15	1.16	1.17	1.17	1.18	1.19	1.19
0.40	1.15	1.16	1.17	1.18	1.19	1.19	1.20	1.21	1.21	1.22
0.30	1.18	1.19	1.20	1.21	1.21	1.22	1.22	1.23	1.23	1.24
0.20	1.21	1.22	1.23	1.23	1.24	1.24	1.25	1.25	1.26	1.26
0.10	1.24	1.25	1.25	1.26	1.26	1.27	1.27	1.27	1.27	1.28
0.00	1.27	1.27	1.28	1.28	1.28	1.28	1.29	1.29	1.29	1.29
–0.10	1.29	1.29	1.29	1.30	1.30	1.30	1.30	1.30	1.30	1.31
–0.20	1.30	1.31	1.31	1.31	1.31	1.31	1.31	1.31	1.32	1.32
–0.30	1.32	1.32	1.32	1.32	1.32	1.32	1.32	1.32	1.32	1.32
–0.40	1.32	1.32	1.32	1.33	1.33	1.33	1.33	1.33	1.33	1.33
–0.50	1.33	1.33	1.33	1.33	1.33	1.33	1.33	1.33	1.33	1.33

When $C_{pk} = C_p$ the average is at the target.

Table 12: The Effective Cost of Production when the cost of rework is 33 percent of the cost of scrap and the average is on the REWORK side of the target

	C_p values							
C_{pk}	*2.5*	*3.0*	*4.0*	*5.0*	*6.0*	*8.0*	*12.0*	*20.0*
20.0	-	-	-	-	-	-	-	1.00
12.0	-	-	-	-	-	-	1.00	1.05
8.0	-	-	-	-	-	1.00	1.04	1.12
6.0	-	-	-	-	1.00	1.02	1.08	1.16
5.0	-	-	-	1.00	1.01	1.05	1.11	1.19
4.0	-	-	1.00	1.01	1.04	1.08	1.15	1.21
3.0	-	1.01	1.02	1.05	1.08	1.13	1.16	1.24
2.5	1.01	1.01	1.05	1.08	1.11	1.16	1.21	1.26
2.00	1.02	1.04	1.09	1.12	1.15	1.19	1.23	1.27
1.90	1.03	1.05	1.09	1.13	1.16	1.19	1.24	1.27
1.80	1.03	1.06	1.10	1.14	1.16	1.20	1.24	1.28
1.70	1.04	1.07	1.11	1.15	1.17	1.21	1.25	1.28
1.60	1.05	1.08	1.12	1.16	1.18	1.21	1.25	1.28
1.50	1.06	1.09	1.13	1.16	1.19	1.22	1.26	1.29
1.40	1.07	1.10	1.14	1.17	1.20	1.23	1.26	1.29
1.30	1.08	1.11	1.15	1.18	1.21	1.23	1.27	1.29
1.20	1.10	1.12	1.17	1.19	1.21	1.24	1.27	1.29
1.10	1.11	1.14	1.18	1.20	1.22	1.25	1.28	1.30
1.00	1.13	1.15	1.19	1.21	1.23	1.26	1.28	1.30
0.90	1.14	1.17	1.20	1.23	1.24	1.26	1.29	1.30
0.80	1.16	1.18	1.22	1.24	1.25	1.27	1.29	1.31
0.70	1.18	1.20	1.23	1.25	1.26	1.28	1.30	1.31
0.60	1.20	1.22	1.24	1.26	1.27	1.29	1.30	1.31
0.50	1.22	1.23	1.26	1.27	1.28	1.29	1.31	1.32
0.40	1.24	1.25	1.27	1.28	1.29	1.30	1.31	1.32
0.30	1.25	1.27	1.28	1.29	1.30	1.31	1.32	1.32
0.20	1.27	1.28	1.29	1.30	1.31	1.31	1.32	1.33
0.10	1.29	1.29	1.30	1.31	1.31	1.32	1.32	1.33
0.00	1.30	1.31	1.31	1.32	1.32	1.32	1.33	1.33
–0.10	1.31	1.32	1.32	1.32	1.32	1.33	1.33	1.33
–0.20	1.32	1.32	1.32	1.33	1.33	1.33	1.33	1.33
–0.30	1.33	1.33	1.33	1.33	1.33	1.33	1.33	1.33
–0.40	1.33	1.33	1.33	1.33	1.33	1.33	1.33	1.33
–0.50	1.33	1.33	1.33	1.33	1.33	1.33	1.33	1.33

Table 13: The Effective Cost of Production when the cost of rework is 20 percent of the cost of scrap and the average is on the SCRAP side of the target

C_{pk}	C_p values 0.10	0.20	0.30	0.40	0.50	0.60	0.70	0.80	0.90	1.00
–0.50	15.23	15.32	15.41	15.49	15.55	15.61	15.65	15.68	15.71	15.73
–0.40	8.94	9.01	9.09	9.17	9.24	9.29	9.33	9.37	9.40	9.42
–0.30	5.66	5.72	5.80	5.87	5.94	5.99	6.04	6.07	6.11	6.13
–0.20	3.87	3.91	3.97	4.04	4.11	4.16	4.21	4.25	4.28	4.31
–0.10	2.83	2.86	2.91	2.97	3.03	3.08	3.13	3.17	3.21	3.24
0.00	2.20	2.22	2.26	2.31	2.37	2.42	2.47	2.51	2.55	2.58
0.10	1.82	1.83	1.85	1.89	1.93	1.98	2.03	2.07	2.11	2.15
0.20	-	1.57	1.58	1.61	1.65	1.69	1.73	1.78	1.82	1.85
0.30	-	-	1.41	1.42	1.45	1.48	1.52	1.57	1.60	1.64
0.40	-	-	-	1.30	1.31	1.34	1.37	1.41	1.45	1.48
0.50	-	-	-	-	1.22	1.24	1.26	1.29	1.33	1.36
0.60	-	-	-	-	-	1.17	1.18	1.21	1.23	1.27
0.70	-	-	-	-	-	-	1.13	1.14	1.17	1.19
0.80	-	-	-	-	-	-	-	1.10	1.12	1.13
0.90	-	-	-	-	-	-	-	-	1.08	1.09
1.00	-	-	-	-	-	-	-	-	-	1.07

Blank spaces correspond to impossible combinations of C_p and C_{pk}.

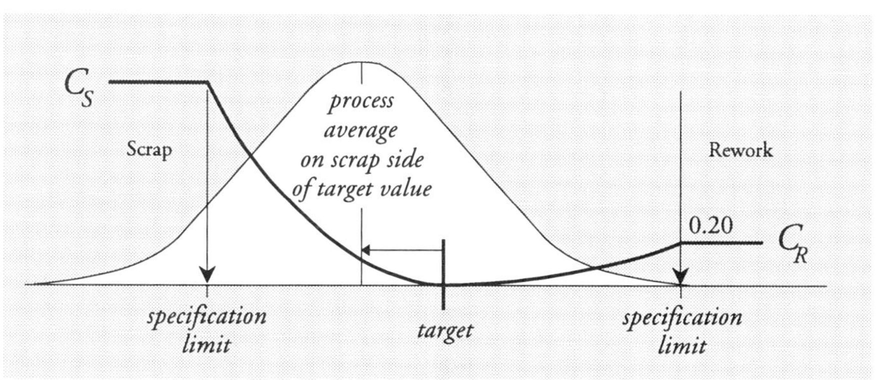

Table 13: Effective Cost of Production
When Cost of Rework is 20% the Cost of Scrap
and the Average is on the Scrap Side of the Target Value

Table 13: The Effective Cost of Production when the cost of rework is 20 percent of the cost of scrap and the average is on the SCRAP side of the target

C_p values

C_{pk}	1.10	1.20	1.30	1.40	1.50	1.60	1.70	1.80	1.90	2.00
–0.50	15.75	15.77	15.78	15.80	15.81	15.82	15.83	15.83	15.84	15.85
–0.40	9.44	9.46	9.48	9.49	9.51	9.52	9.53	9.53	9.54	9.55
–0.30	6.16	6.18	6.19	6.21	6.22	6.23	6.24	6.25	6.26	6.27
–0.20	4.33	4.36	4.38	4.39	4.41	4.42	4.43	4.44	4.45	4.46
–0.10	3.27	3.29	3.31	3.33	3.35	3.36	3.37	3.39	3.40	3.41
0.00	2.61	2.63	2.66	2.68	2.70	2.71	2.73	2.74	2.75	2.76
0.10	2.18	2.21	2.23	2.25	2.27	2.29	2.31	2.32	2.34	2.35
0.20	1.89	1.92	1.94	1.97	1.99	2.01	2.03	2.04	2.06	2.07
0.30	1.68	1.71	1.74	1.76	1.78	1.81	1.83	1.84	1.86	1.87
0.40	1.52	1.55	1.58	1.61	1.63	1.66	1.68	1.70	1.71	1.73
0.50	1.40	1.43	1.46	1.49	1.51	1.54	1.56	1.58	1.60	1.62
0.60	1.30	1.33	1.36	1.39	1.42	1.44	1.47	1.49	1.51	1.53
0.70	1.22	1.25	1.28	1.31	1.34	1.36	1.39	1.41	1.43	1.46
0.80	1.16	1.19	1.21	1.24	1.27	1.29	1.32	1.34	1.37	1.39
0.90	1.11	1.13	1.16	1.18	1.21	1.24	1.26	1.28	1.31	1.33
1.00	1.08	1.09	1.11	1.14	1.16	1.18	1.21	1.23	1.26	1.28
1.10	1.05	1.07	1.08	1.10	1.12	1.14	1.16	1.19	1.21	1.23
1.20	-	1.05	1.06	1.07	1.09	1.10	1.12	1.15	1.17	1.19
1.30	-	-	1.04	1.05	1.06	1.08	1.09	1.11	1.13	1.15
1.40	-	-	-	1.03	1.04	1.05	1.07	1.08	1.10	1.12
1.50	-	-	-	-	1.03	1.04	1.05	1.06	1.07	1.09
1.60	-	-	-	-	-	1.03	1.03	1.04	1.05	1.07
1.70	-	-	-	-	-	-	1.02	1.03	1.04	1.05
1.80	-	-	-	-	-	-	-	1.02	1.03	1.03
1.90	-	-	-	-	-	-	-	-	1.02	1.02
2.00	-	-	-	-	-	-	-	-	-	1.02

When $C_{pk} = C_p$ the average is at the target.

Table 13: The Effective Cost of Production when the cost of rework is 20 percent of the cost of scrap and the average is on the SCRAP side of the target

C_{pk}	C_p values 2.5	3.0	4.0	5.0	6.0	8.0	12.0	20.0
–0.50	15.91	15.93	15.95	15.97	15.98	15.99	16.00	16.01
–0.40	9.59	9.61	9.64	9.65	9.66	9.68	9.69	9.70
–0.30	6.31	6.33	6.36	6.38	6.39	6.40	6.42	6.43
–0.20	4.50	4.52	4.56	4.57	4.59	4.60	4.62	4.63
–0.10	3.45	3.48	3.51	3.53	3.55	3.56	3.58	3.60
0.00	2.81	2.84	2.88	2.90	2.92	2.94	2.96	2.98
0.10	2.40	2.43	2.48	2.50	2.52	2.55	2.57	2.59
0.20	2.12	2.16	2.21	2.24	2.27	2.29	2.32	2.34
0.30	1.94	1.98	2.04	2.07	2.10	2.13	2.16	2.19
0.40	1.80	1.85	1.91	1.95	1.98	2.02	2.05	2.08
0.50	1.70	1.75	1.82	1.87	1.90	1.94	1.98	2.02
0.60	1.61	1.67	1.75	1.81	1.84	1.89	1.94	1.98
0.70	1.54	1.61	1.70	1.76	1.80	1.85	1.90	1.95
0.80	1.48	1.55	1.65	1.72	1.76	1.82	1.88	1.93
0.90	1.43	1.50	1.61	1.68	1.73	1.79	1.86	1.92
1.00	1.38	1.46	1.57	1.65	1.70	1.77	1.84	1.90
1.10	1.33	1.41	1.53	1.61	1.67	1.75	1.83	1.89
1.20	1.29	1.37	1.50	1.58	1.64	1.72	1.81	1.88
1.30	1.25	1.33	1.46	1.55	1.62	1.70	1.80	1.87
1.40	1.21	1.30	1.43	1.52	1.59	1.68	1.78	1.87
1.50	1.18	1.26	1.40	1.49	1.57	1.66	1.77	1.86
1.60	1.15	1.23	1.37	1.47	1.54	1.64	1.75	1.85
1.70	1.12	1.20	1.34	1.44	1.52	1.62	1.74	1.84
1.80	1.10	1.17	1.31	1.41	1.49	1.60	1.72	1.83
1.90	1.08	1.15	1.28	1.39	1.47	1.58	1.71	1.82
2.00	1.06	1.12	1.26	1.36	1.45	1.56	1.70	1.81
2.5	1.01	1.04	1.15	1.25	1.34	1.47	1.63	1.77
3.0	-	1.01	1.07	1.16	1.25	1.39	1.56	1.72
4.0	-	-	1.00	1.04	1.11	1.25	1.45	1.64
5.0	-	-	-	1.00	1.03	1.14	1.34	1.56
6.0	-	-	-	-	1.00	1.06	1.25	1.49
8.0	-	-	-	-	-	1.00	1.11	1.36
12.0	-	-	-	-	-	-	1.00	1.16
20.0	-	-	-	-	-	-	-	1.00

Table 14: The Effective Cost of Production when the cost of rework is 20 percent of the cost of scrap and the average is on the REWORK side of the target

	C_p values									
C_{pk}	**0.10**	**0.20**	**0.30**	**0.40**	**0.50**	**0.60**	**0.70**	**0.80**	**0.90**	**1.00**
1.00	-	-	-	-	-	-	-	-	-	1.07
0.90	-	-	-	-	-	-	-	-	1.08	1.05
0.80	-	-	-	-	-	-	-	1.10	1.06	1.05
0.70	-	-	-	-	-	-	1.13	1.08	1.06	1.05
0.60	-	-	-	-	-	1.17	1.10	1.07	1.06	1.06
0.50	-	-	-	-	1.22	1.13	1.09	1.07	1.07	1.07
0.40	-	-	-	1.30	1.17	1.11	1.09	1.08	1.08	1.09
0.30	-	-	1.41	1.22	1.14	1.11	1.10	1.10	1.10	1.11
0.20	-	1.57	1.30	1.18	1.13	1.11	1.11	1.11	1.12	1.12
0.10	1.82	1.41	1.24	1.16	1.13	1.13	1.13	1.13	1.14	1.14
0.00	1.58	1.31	1.20	1.16	1.15	1.14	1.15	1.15	1.15	1.16
–0.10	1.42	1.25	1.19	1.16	1.16	1.16	1.16	1.17	1.17	1.17
–0.20	1.33	1.22	1.18	1.17	1.17	1.17	1.18	1.18	1.18	1.18
–0.30	1.27	1.21	1.19	1.18	1.18	1.18	1.19	1.19	1.19	1.19
–0.40	1.23	1.20	1.19	1.19	1.19	1.19	1.19	1.19	1.19	1.19
–0.50	1.22	1.20	1.19	1.19	1.19	1.19	1.20	1.20	1.20	1.20

Blank spaces correspond to impossible combinations of C_p and C_{pk}.

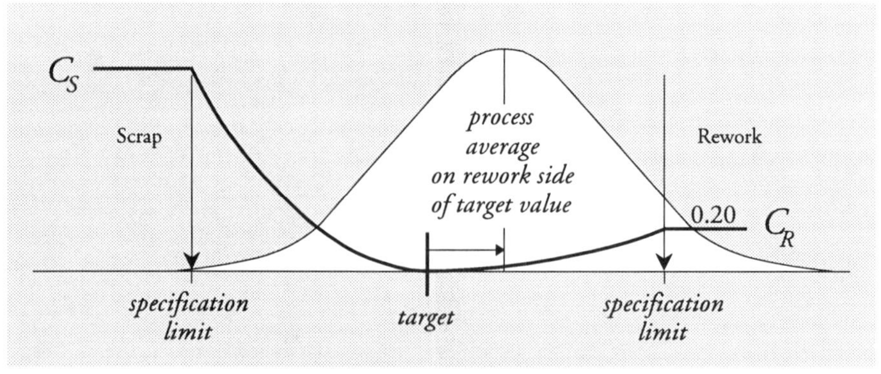

Table 14: Effective Cost of Production When Cost of Rework is 20% the Cost of Scrap and the Average is on the Rework Side of the Target Value

Table 14: The Effective Cost of Production when the cost of rework is 20 percent of the cost of scrap and the average is on the REWORK side of the target

C_{pk}	C_p values: 1.10	1.20	1.30	1.40	1.50	1.60	1.70	1.80	1.90	2.00
2.00	-	-	-	-	-	-	-	-	-	1.02
1.90	-	-	-	-	-	-	-	-	1.02	1.01
1.80	-	-	-	-	-	-	-	1.02	1.01	1.01
1.70	-	-	-	-	-	-	1.02	1.02	1.01	1.01
1.60	-	-	-	-	-	1.03	1.02	1.01	1.01	1.01
1.50	-	-	-	-	1.03	1.02	1.02	1.01	1.02	1.02
1.40	-	-	-	1.03	1.02	1.02	1.02	1.02	1.02	1.02
1.30	-	-	1.04	1.03	1.02	1.02	1.02	1.02	1.03	1.03
1.20	-	1.05	1.03	1.02	1.02	1.02	1.03	1.03	1.03	1.04
1.10	1.05	1.04	1.03	1.02	1.03	1.03	1.03	1.04	1.04	1.05
1.00	1.04	1.03	1.03	1.03	1.03	1.04	1.04	1.05	1.05	1.06
0.90	1.04	1.03	1.03	1.04	1.04	1.05	1.05	1.06	1.06	1.07
0.80	1.04	1.04	1.04	1.05	1.05	1.06	1.06	1.07	1.07	1.08
0.70	1.05	1.05	1.06	1.06	1.07	1.07	1.08	1.08	1.09	1.09
0.60	1.06	1.06	1.07	1.08	1.08	1.09	1.09	1.09	1.10	1.10
0.50	1.07	1.08	1.09	1.09	1.10	1.10	1.10	1.11	1.11	1.12
0.40	1.09	1.10	1.10	1.11	1.11	1.12	1.12	1.12	1.13	1.13
0.30	1.11	1.12	1.12	1.12	1.13	1.13	1.13	1.14	1.14	1.14
0.20	1.13	1.13	1.14	1.14	1.14	1.15	1.15	1.15	1.15	1.16
0.10	1.15	1.15	1.15	1.15	1.16	1.16	1.16	1.16	1.16	1.17
0.00	1.16	1.16	1.17	1.17	1.17	1.17	1.17	1.17	1.17	1.18
–0.10	1.17	1.17	1.18	1.18	1.18	1.18	1.18	1.18	1.18	1.18
–0.20	1.18	1.18	1.18	1.19	1.19	1.19	1.19	1.19	1.19	1.19
–0.30	1.19	1.19	1.19	1.19	1.19	1.19	1.19	1.19	1.19	1.19
–0.40	1.19	1.19	1.19	1.20	1.20	1.20	1.20	1.20	1.20	1.20
–0.50	1.20	1.20	1.20	1.20	1.20	1.20	1.20	1.20	1.20	1.20

When $C_{pk} = C_p$ the average is at the target.

Table 14: The Effective Cost of Production when the cost of rework is 20 percent of the cost of scrap and the average is on the REWORK side of the target

	C_p values							
C_{pk}	**2.5**	**3.0**	**4.0**	**5.0**	**6.0**	**8.0**	**12.0**	**20.0**
20.0	-	-	-	-	-	-	-	1.00
12.0	-	-	-	-	-	-	1.00	1.03
8.0	-	-	-	-	-	1.00	1.02	1.17
6.0	-	-	-	-	1.00	1.01	1.05	1.10
5.0	-	-	-	1.00	1.01	1.03	1.07	1.11
4.0	-	-	1.00	1.01	1.02	1.05	1.09	1.13
3.0	-	1.01	1.01	1.03	1.05	1.08	1.11	1.14
2.5	1.01	1.01	1.03	1.05	1.07	1.09	1.13	1.15
2.00	1.01	1.02	1.05	1.07	1.09	1.11	1.14	1.16
1.90	1.02	1.03	1.06	1.08	1.09	1.12	1.14	1.16
1.80	1.02	1.03	1.06	1.08	1.10	1.12	1.14	1.17
1.70	1.02	1.04	1.07	1.09	1.10	1.12	1.15	1.17
1.60	1.03	1.05	1.07	1.09	1.11	1.13	1.15	1.17
1.50	1.04	1.05	1.08	1.10	1.11	1.13	1.15	1.17
1.40	1.04	1.06	1.09	1.10	1.12	1.14	1.16	1.17
1.30	1.05	1.07	1.09	1.11	1.12	1.14	1.16	1.17
1.20	1.06	1.07	1.10	1.12	1.13	1.14	1.16	1.18
1.10	1.07	1.08	1.11	1.12	1.13	1.15	1.17	1.18
1.00	1.08	1.09	1.11	1.13	1.14	1.15	1.17	1.18
0.90	1.09	1.10	1.12	1.14	1.15	1.16	1.17	1.18
0.80	1.10	1.11	1.13	1.14	1.15	1.16	1.17	1.18
0.70	1.11	1.12	1.14	1.15	1.16	1.17	1.18	1.19
0.60	1.12	1.13	1.15	1.16	1.16	1.17	1.18	1.19
0.50	1.13	1.14	1.15	1.16	1.17	1.18	1.18	1.19
0.40	1.14	1.15	1.16	1.17	1.17	1.18	1.19	1.19
0.30	1.15	1.16	1.17	1.17	1.18	1.18	1.19	1.19
0.20	1.16	1.17	1.18	1.18	1.18	1.19	1.19	1.20
0.10	1.17	1.18	1.18	1.19	1.19	1.19	1.19	1.20
0.00	1.18	1.18	1.19	1.19	1.19	1.19	1.20	1.20
–0.10	1.19	1.19	1.19	1.19	1.19	1.20	1.20	1.20
–0.20	1.19	1.19	1.19	1.20	1.20	1.20	1.20	1.20
–0.30	1.20	1.20	1.20	1.20	1.20	1.20	1.20	1.20
–0.40	1.20	1.20	1.20	1.20	1.20	1.20	1.20	1.20
–0.50	1.20	1.20	1.20	1.20	1.20	1.20	1.20	1.20

Table 15: The Effective Cost of Production when the cost of rework is 10 percent of the cost of scrap and the average is on the SCRAP side of the target

	C_p values									
C_{pk}	**0.10**	**0.20**	**0.30**	**0.40**	**0.50**	**0.60**	**0.70**	**0.80**	**0.90**	**1.00**
–0.50	15.20	15.31	15.41	15.49	15.55	15.61	15.65	15.68	15.71	15.73
–0.40	8.90	9.00	9.09	9.17	9.24	9.29	9.33	9.37	9.40	9.42
–0.30	5.62	5.71	5.79	5.87	5.94	5.99	6.04	6.07	6.11	6.13
–0.20	3.82	3.89	3.97	4.04	4.10	4.16	4.21	4.25	4.28	4.31
–0.10	2.77	2.83	2.90	2.97	3.03	3.08	3.13	3.17	3.21	3.24
0.00	2.14	2.19	2.25	2.31	2.36	2.42	2.47	2.51	2.55	2.58
0.10	1.75	1.79	1.83	1.88	1.93	1.98	2.03	2.07	2.11	2.15
0.20	-	1.53	1.56	1.60	1.64	1.69	1.73	1.78	1.82	1.85
0.30	-	-	1.38	1.40	1.44	1.48	1.52	1.56	1.60	1.64
0.40	-	-	-	1.27	1.30	1.33	1.37	1.41	1.45	1.48
0.50	-	-	-	-	1.20	1.23	1.26	1.29	1.33	1.36
0.60	-	-	-	-	-	1.15	1.18	1.20	1.23	1.27
0.70	-	-	-	-	-	-	1.12	1.14	1.16	1.19
0.80	-	-	-	-	-	-	-	1.09	1.11	1.13
0.90	-	-	-	-	-	-	-	-	1.07	1.09
1.00	-	-	-	-	-	-	-	-	-	1.06

Blank spaces correspond to impossible combinations of C_p and C_{pk}.

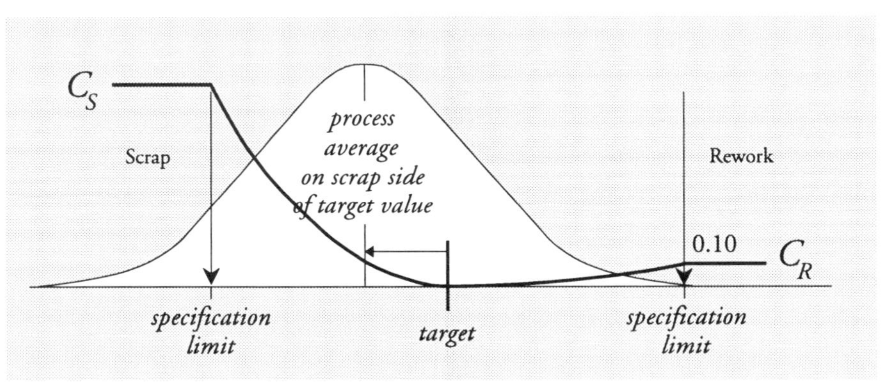

Table 15: Effective Cost of Production
When Cost of Rework is 10% the Cost of Scrap
and the Average is on the Scrap Side of the Target Value

Table 15: The Effective Cost of Production when the cost of rework is 10 percent of the cost of scrap and the average is on the SCRAP side of the target

C_{pk}	C_p values 1.10	1.20	1.30	1.40	1.50	1.60	1.70	1.80	1.90	2.00
–0.50	15.75	15.77	15.78	15.80	15.81	15.82	15.83	15.83	15.84	15.85
–0.40	9.44	9.46	9.48	9.49	9.51	9.52	9.53	9.53	9.54	9.55
–0.30	6.16	6.18	6.19	6.21	6.22	6.23	6.24	6.25	6.26	6.27
–0.20	4.33	4.36	4.38	4.39	4.41	4.42	4.43	4.44	4.45	4.46
–0.10	3.27	3.29	3.31	3.33	3.35	3.36	3.37	3.39	3.40	3.41
0.00	2.61	2.63	2.66	2.68	2.70	2.71	2.73	2.74	2.75	2.76
0.10	2.18	2.21	2.23	2.25	2.27	2.29	2.31	2.32	2.34	2.35
0.20	1.89	1.92	1.94	1.97	1.99	2.01	2.03	2.04	2.06	2.07
0.30	1.68	1.71	1.74	1.76	1.78	1.81	1.83	1.84	1.86	1.87
0.40	1.52	1.55	1.58	1.61	1.63	1.66	1.68	1.70	1.71	1.73
0.50	1.39	1.43	1.46	1.49	1.51	1.54	1.56	1.58	1.60	1.62
0.60	1.30	1.33	1.36	1.39	1.42	1.44	1.47	1.49	1.51	1.53
0.70	1.22	1.25	1.28	1.31	1.34	1.36	1.39	1.41	1.43	1.46
0.80	1.16	1.18	1.21	1.24	1.27	1.29	1.32	1.34	1.37	1.39
0.90	1.11	1.13	1.16	1.18	1.21	1.23	1.26	1.28	1.31	1.33
1.00	1.07	1.09	1.11	1.14	1.16	1.18	1.21	1.23	1.26	1.28
1.10	1.05	1.06	1.08	1.10	1.12	1.14	1.16	1.19	1.21	1.23
1.20	-	1.04	1.05	1.07	1.09	1.10	1.12	1.15	1.17	1.19
1.30	-	-	1.04	1.05	1.06	1.07	1.09	1.11	1.13	1.15
1.40	-	-	-	1.03	1.04	1.05	1.07	1.08	1.10	1.12
1.50	-	-	-	-	1.03	1.04	1.05	1.06	1.07	1.09
1.60	-	-	-	-	-	1.02	1.03	1.04	1.05	1.07
1.70	-	-	-	-	-	-	1.02	1.03	1.04	1.05
1.80	-	-	-	-	-	-	-	1.02	1.03	1.03
1.90	-	-	-	-	-	-	-	-	1.02	1.02
2.00	-	-	-	-	-	-	-	-	-	1.02

When $C_{pk} = C_p$ the average is at the target.

Table 15: The Effective Cost of Production when the cost of rework is 10 percent of the cost of scrap and the average is on the SCRAP side of the target

C_{pk}	C_p values 2.5	3.0	4.0	5.0	6.0	8.0	12.0	20.0
–0.50	15.91	15.93	15.95	15.97	15.98	15.99	16.00	16.01
–0.40	9.59	9.61	9.64	9.65	9.66	9.68	9.69	9.70
–0.30	6.31	6.33	6.36	6.38	6.39	6.40	6.42	6.43
–0.20	4.50	4.52	4.56	4.57	4.59	4.60	4.62	4.63
–0.10	3.45	3.48	3.51	3.53	3.55	3.56	3.58	3.60
0.00	2.81	2.84	2.88	2.90	2.92	2.94	2.96	2.98
0.10	2.40	2.43	2.48	2.50	2.52	2.55	2.57	2.59
0.20	2.12	2.16	2.21	2.24	2.27	2.29	2.32	2.34
0.30	1.94	1.98	2.04	2.07	2.10	2.13	2.16	2.19
0.40	1.80	1.85	1.91	1.95	1.98	2.02	2.05	2.08
0.50	1.70	1.75	1.82	1.87	1.90	1.94	1.98	2.02
0.60	1.61	1.67	1.75	1.81	1.84	1.89	1.94	1.98
0.70	1.54	1.61	1.70	1.76	1.80	1.85	1.90	1.95
0.80	1.48	1.55	1.65	1.72	1.76	1.82	1.88	1.93
0.90	1.43	1.50	1.61	1.68	1.73	1.79	1.86	1.92
1.00	1.38	1.46	1.57	1.65	1.70	1.77	1.84	1.90
1.10	1.33	1.41	1.53	1.61	1.67	1.75	1.83	1.89
1.20	1.29	1.37	1.50	1.58	1.64	1.72	1.81	1.88
1.30	1.25	1.33	1.46	1.55	1.62	1.70	1.80	1.87
1.40	1.21	1.30	1.43	1.52	1.59	1.68	1.78	1.87
1.50	1.18	1.26	1.40	1.49	1.57	1.66	1.77	1.86
1.60	1.15	1.23	1.37	1.47	1.54	1.64	1.75	1.85
1.70	1.12	1.20	1.34	1.44	1.52	1.62	1.74	1.84
1.80	1.10	1.17	1.31	1.41	1.49	1.60	1.72	1.83
1.90	1.08	1.15	1.28	1.39	1.47	1.58	1.71	1.82
2.00	1.06	1.12	1.26	1.36	1.45	1.56	1.70	1.81
2.5	1.01	1.04	1.15	1.25	1.34	1.47	1.63	1.77
3.0	-	1.01	1.07	1.16	1.25	1.39	1.56	1.72
4.0	-	-	1.00	1.04	1.11	1.25	1.45	1.64
5.0	-	-	-	1.00	1.03	1.14	1.34	1.56
6.0	-	-	-	-	1.00	1.06	1.25	1.49
8.0	-	-	-	-	-	1.00	1.11	1.36
12.0	-	-	-	-	-	-	1.00	1.16
20.0	-	-	-	-	-	-	-	1.00

Table 16: The Effective Cost of Production when the cost of rework is 10 percent of the cost of scrap and the average is on the REWORK side of the target

C_{pk}	C_p values 0.10	0.20	0.30	0.40	0.50	0.60	0.70	0.80	0.90	1.00
1.00	-	-	-	-	-	-	-	-	-	1.06
0.90	-	-	-	-	-	-	-	-	1.07	1.04
0.80	-	-	-	-	-	-	-	1.09	1.05	1.03
0.70	-	-	-	-	-	-	1.12	1.06	1.04	1.03
0.60	-	-	-	-	-	1.15	1.08	1.05	1.04	1.03
0.50	-	-	-	-	1.20	1.11	1.06	1.04	1.04	1.04
0.40	-	-	-	1.27	1.14	1.08	1.05	1.04	1.04	1.04
0.30	-	-	1.38	1.19	1.10	1.07	1.05	1.05	1.05	1.05
0.20	-	1.52	1.26	1.14	1.08	1.06	1.06	1.06	1.06	1.06
0.10	1.75	1.36	1.18	1.11	1.08	1.07	1.07	1.07	1.07	1.07
0.00	1.50	1.25	1.14	1.09	1.08	1.07	1.07	1.08	1.08	1.08
–0.10	1.34	1.18	1.11	1.09	1.08	1.08	1.08	1.08	1.08	1.09
–0.20	1.24	1.14	1.10	1.09	1.09	1.09	1.09	1.09	1.09	1.09
–0.30	1.18	1.12	1.10	1.09	1.09	1.09	1.09	1.09	1.09	1.09
–0.40	1.14	1.11	1.10	1.09	1.09	1.10	1.10	1.10	1.10	1.10
–0.50	1.12	1.10	1.10	1.10	1.10	1.10	1.10	1.10	1.10	1.10

Blank spaces correspond to impossible combinations of C_p and C_{pk}.

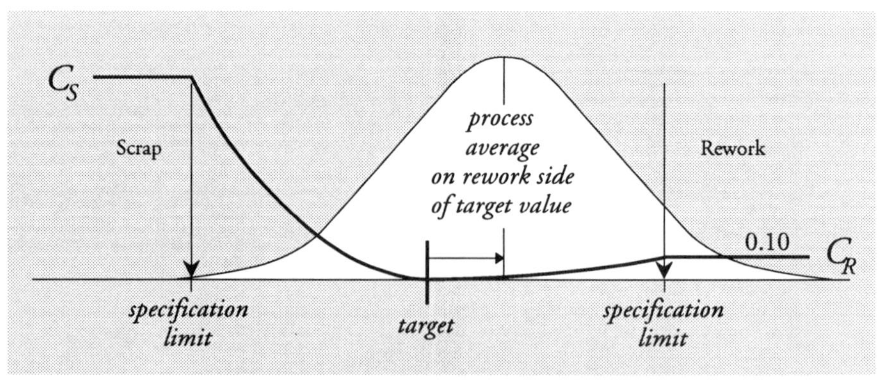

Table 16: Effective Cost of Production When Cost of Rework is 10% the Cost of Scrap and the Average is on the Rework Side of the Target Value

Table 16: The Effective Cost of Production when the cost of rework is 10 percent of the cost of scrap and the average is on the REWORK side of the target

	C_p *values*									
C_{pk}	***1.10***	***1.20***	***1.30***	***1.40***	***1.50***	***1.60***	***1.70***	***1.80***	***1.90***	***2.00***
2.00	-	-	-	-	-	-	-	-	-	1.02
1.90	-	-	-	-	-	-	-	-	1.02	1.01
1.80	-	-	-	-	-	-	-	1.02	1.01	1.01
1.70	-	-	-	-	-	-	1.02	1.01	1.01	1.01
1.60	-	-	-	-	-	1.02	1.01	1.01	1.01	1.01
1.50	-	-	-	-	1.03	1.02	1.01	1.01	1.01	1.01
1.40	-	-	-	1.03	1.02	1.01	1.01	1.01	1.01	1.01
1.30	-	-	1.04	1.02	1.01	1.01	1.01	1.01	1.01	1.02
1.20	-	1.04	1.03	1.02	1.01	1.01	1.01	1.01	1.02	1.02
1.10	1.05	1.03	1.02	1.02	1.01	1.01	1.02	1.02	1.02	1.02
1.00	1.03	1.02	1.02	1.02	1.02	1.02	1.02	1.02	1.03	1.03
0.90	1.03	1.02	1.02	1.02	1.02	1.02	1.03	1.03	1.03	1.03
0.80	1.02	1.02	1.02	1.02	1.03	1.03	1.03	1.03	1.04	1.04
0.70	1.03	1.03	1.03	1.03	1.03	1.04	1.04	1.04	1.04	1.04
0.60	1.03	1.03	1.04	1.04	1.04	1.04	1.05	1.05	1.05	1.05
0.50	1.04	1.04	1.04	1.05	1.05	1.05	1.05	1.05	1.06	1.06
0.40	1.05	1.05	1.05	1.05	1.06	1.06	1.06	1.06	1.06	1.06
0.30	1.06	1.06	1.06	1.06	1.06	1.07	1.07	1.07	1.07	1.07
0.20	1.06	1.07	1.07	1.07	1.07	1.07	1.07	1.08	1.08	1.08
0.10	1.07	1.07	1.08	1.08	1.08	1.08	1.08	1.08	1.08	1.08
0.00	1.08	1.08	1.08	1.08	1.08	1.09	1.09	1.09	1.09	1.09
–0.10	1.09	1.09	1.09	1.09	1.09	1.09	1.09	1.09	1.09	1.09
–0.20	1.09	1.09	1.09	1.09	1.09	1.09	1.09	1.09	1.09	1.09
–0.30	1.09	1.10	1.10	1.10	1.10	1.10	1.10	1.10	1.10	1.10
–0.40	1.10	1.10	1.10	1.10	1.10	1.10	1.10	1.10	1.10	1.10
–0.50	1.10	1.10	1.10	1.10	1.10	1.10	1.10	1.10	1.10	1.10

When $C_{pk} = C_p$ the average is at the target.

Table 16: The Effective Cost of Production when the cost of rework is 10 percent of the cost of scrap and the average is on the REWORK side of the target

C_{pk}	C_p values: 2.5	3.0	4.0	5.0	6.0	8.0	12.0	20.0
20.0	-	-	-	-	-	-	-	1.00
12.0	-	-	-	-	-	-	1.00	1.02
8.0	-	-	-	-	-	1.00	1.01	1.04
6.0	-	-	-	-	1.00	1.01	1.03	1.05
5.0	-	-	-	1.00	1.00	1.01	1.03	1.06
4.0	-	-	1.00	1.00	1.01	1.03	1.04	1.06
3.0	-	1.01	1.01	1.02	1.03	1.04	1.06	1.07
2.5	1.01	1.00	1.01	1.03	1.03	1.05	1.06	1.08
2.00	1.01	1.01	1.03	1.04	1.04	1.06	1.07	1.08
1.90	1.01	1.01	1.03	1.04	1.05	1.06	1.07	1.08
1.80	1.01	1.02	1.03	1.04	1.05	1.06	1.07	1.08
1.70	1.01	1.02	1.03	1.04	1.05	1.06	1.07	1.08
1.60	1.01	1.02	1.04	1.05	1.05	1.06	1.08	1.08
1.50	1.02	1.03	1.04	1.05	1.06	1.07	1.08	1.09
1.40	1.02	1.03	1.04	1.05	1.06	1.07	1.08	1.09
1.30	1.02	1.03	1.05	1.06	1.06	1.07	1.08	1.09
1.20	1.03	1.04	1.05	1.06	1.06	1.07	1.08	1.09
1.10	1.03	1.04	1.05	1.06	1.07	1.07	1.08	1.09
1.00	1.04	1.05	1.06	1.06	1.07	1.08	1.08	1.09
0.90	1.04	1.05	1.06	1.07	1.07	1.08	1.09	1.09
0.80	1.05	1.05	1.06	1.07	1.08	1.08	1.09	1.09
0.70	1.05	1.06	1.07	1.07	1.08	1.08	1.09	1.09
0.60	1.06	1.06	1.07	1.08	1.08	1.09	1.09	1.09
0.50	1.06	1.07	1.08	1.08	1.08	1.09	1.09	1.10
0.40	1.07	1.08	1.08	1.08	1.09	1.09	1.09	1.10
0.30	1.08	1.08	1.08	1.09	1.09	1.09	1.09	1.10
0.20	1.08	1.08	1.09	1.09	1.09	1.09	1.10	1.10
0.10	1.09	1.09	1.09	1.09	1.09	1.10	1.10	1.10
0.00	1.09	1.09	1.09	1.10	1.10	1.10	1.10	1.10
–0.10	1.09	1.09	1.10	1.10	1.10	1.10	1.10	1.10
–0.20	1.10	1.10	1.10	1.10	1.10	1.10	1.10	1.10
–0.30	1.10	1.10	1.10	1.10	1.10	1.10	1.10	1.10
–0.40	1.10	1.10	1.10	1.10	1.10	1.10	1.10	1.10
–0.50	1.10	1.10	1.10	1.10	1.10	1.10	1.10	1.10

Table 17: The Effective Cost of Production when the cost of rework is ONE percent of the cost of scrap and the average is on the SCRAP side of the target

C_{pk}	C_p values 0.10	0.20	0.30	0.40	0.50	0.60	0.70	0.80	0.90	1.00
–0.50	15.17	15.30	15.41	15.49	15.55	15.61	15.65	15.68	15.71	15.73
–0.40	8.86	8.99	9.09	9.17	9.24	9.29	9.33	9.37	9.40	9.42
–0.30	5.58	5.69	5.79	5.87	5.93	5.99	6.04	6.07	6.11	6.13
–0.20	3.77	3.87	3.96	4.04	4.10	4.16	4.21	4.25	4.28	4.31
–0.10	2.72	2.81	2.89	2.96	3.03	3.08	3.13	3.17	3.21	3.24
0.00	2.09	2.16	2.23	2.30	2.36	2.42	2.47	2.51	2.55	2.58
0.10	1.69	1.75	1.81	1.87	1.93	1.98	2.03	2.07	2.11	2.15
0.20	-	1.48	1.53	1.58	1.64	1.69	1.73	1.78	1.82	1.85
0.30	-	-	1.34	1.39	1.43	1.48	1.52	1.56	1.60	1.64
0.40	-	-	-	1.25	1.29	1.33	1.37	1.41	1.45	1.48
0.50	-	-	-	-	1.19	1.22	1.25	1.29	1.33	1.36
0.60	-	-	-	-	-	1.14	1.17	1.20	1.23	1.26
0.70	-	-	-	-	-	-	1.11	1.13	1.16	1.19
0.80	-	-	-	-	-	-	-	1.09	1.11	1.13
0.90	-	-	-	-	-	-	-	-	1.07	1.09
1.00	-	-	-	-	-	-	-	-	-	1.06

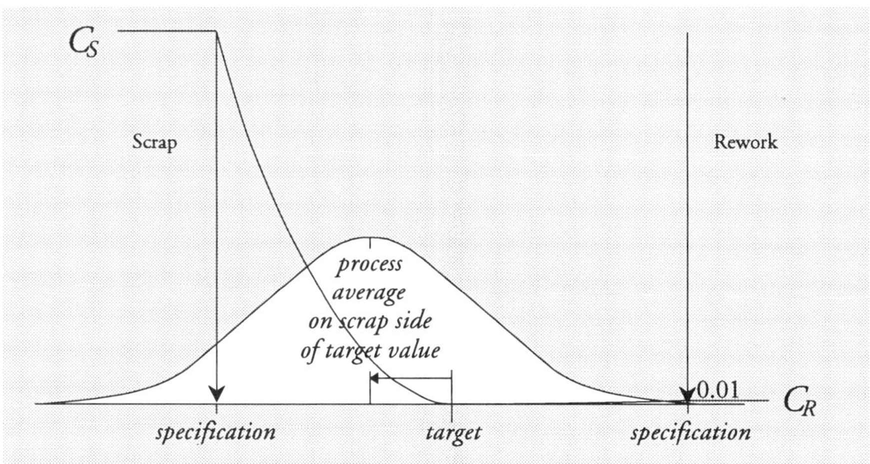

Table 17: Effective Cost of Production
When Cost of Rework is 1% the Cost of Scrap
and the Average is on the Scrap Side of the Target Value

Table 17: The Effective Cost of Production when the cost of rework is ONE percent of the cost of scrap and the average is on the SCRAP side of the target

C_{pk}	C_p values 1.10	1.20	1.30	1.40	1.50	1.60	1.70	1.80	1.90	2.00
–0.50	15.75	15.77	15.78	15.80	15.81	15.82	15.83	15.83	15.84	15.85
–0.40	9.44	9.46	9.48	9.49	9.51	9.52	9.53	9.53	9.54	9.55
–0.30	6.16	6.18	6.19	6.21	6.22	6.23	6.24	6.25	6.26	6.27
–0.20	4.33	4.36	4.38	4.39	4.41	4.42	4.43	4.44	4.45	4.46
–0.10	3.27	3.29	3.31	3.33	3.35	3.36	3.37	3.39	3.40	3.41
0.00	2.61	2.63	2.66	2.68	2.70	2.71	2.73	2.74	2.75	2.76
0.10	2.18	2.21	2.23	2.25	2.27	2.29	2.31	2.32	2.34	2.35
0.20	1.89	1.92	1.94	1.97	1.99	2.01	2.03	2.04	2.06	2.07
0.30	1.68	1.71	1.74	1.76	1.78	1.81	1.83	1.84	1.86	1.87
0.40	1.52	1.55	1.58	1.61	1.63	1.66	1.68	1.70	1.71	1.73
0.50	1.39	1.43	1.46	1.49	1.51	1.54	1.56	1.58	1.60	1.62
0.60	1.30	1.33	1.36	1.39	1.42	1.44	1.47	1.49	1.51	1.53
0.70	1.22	1.25	1.28	1.31	1.34	1.36	1.39	1.41	1.43	1.46
0.80	1.16	1.18	1.21	1.24	1.27	1.29	1.32	1.34	1.37	1.39
0.90	1.11	1.13	1.16	1.18	1.21	1.23	1.26	1.28	1.31	1.33
1.00	1.07	1.09	1.11	1.14	1.16	1.18	1.21	1.23	1.26	1.28
1.10	1.05	1.06	1.08	1.10	1.12	1.14	1.16	1.19	1.21	1.23
1.20	-	1.04	1.05	1.07	1.08	1.10	1.12	1.15	1.17	1.19
1.30	-	-	1.03	1.04	1.06	1.07	1.09	1.11	1.13	1.15
1.40	-	-	-	1.03	1.04	1.05	1.07	1.08	1.10	1.12
1.50	-	-	-	-	1.02	1.03	1.05	1.06	1.07	1.09
1.60	-	-	-	-	-	1.02	1.03	1.04	1.05	1.07
1.70	-	-	-	-	-	-	1.02	1.03	1.04	1.05
1.80	-	-	-	-	-	-	-	1.02	1.02	1.03
1.90	-	-	-	-	-	-	-	-	1.02	1.02
2.00	-	-	-	-	-	-	-	-	-	1.01

Blank spaces correspond to impossible combinations of C_p and C_{pk}.

When $C_{pk} = C_p$ the average is at the target.

Table 17: The Effective Cost of Production when the cost of rework is ONE percent of the cost of scrap and the average is on the SCRAP side of the target

	C_p values							
C_{pk}	*2.5*	*3.0*	*4.0*	*5.0*	*6.0*	*8.0*	*12.0*	*20.0*
–0.50	15.91	15.93	15.95	15.97	15.98	15.99	16.00	16.01
–0.40	9.59	9.61	9.64	9.65	9.66	9.68	9.69	9.70
–0.30	6.31	6.33	6.36	6.38	6.39	6.40	6.42	6.43
–0.20	4.50	4.52	4.56	4.57	4.59	4.60	4.62	4.63
–0.10	3.45	3.48	3.51	3.53	3.55	3.56	3.58	3.60
0.00	2.81	2.84	2.88	2.90	2.92	2.94	2.96	2.98
0.10	2.40	2.43	2.48	2.50	2.52	2.55	2.57	2.59
0.20	2.12	2.16	2.21	2.24	2.27	2.29	2.32	2.34
0.30	1.94	1.98	2.04	2.07	2.10	2.13	2.16	2.19
0.40	1.80	1.85	1.91	1.95	1.98	2.02	2.05	2.08
0.50	1.70	1.75	1.82	1.87	1.90	1.94	1.98	2.02
0.60	1.61	1.67	1.75	1.81	1.84	1.89	1.94	1.98
0.70	1.54	1.61	1.70	1.76	1.80	1.85	1.90	1.95
0.80	1.48	1.55	1.65	1.72	1.76	1.82	1.88	1.93
0.90	1.43	1.50	1.61	1.68	1.73	1.79	1.86	1.92
1.00	1.38	1.46	1.57	1.65	1.70	1.77	1.84	1.90
1.10	1.33	1.41	1.53	1.61	1.67	1.75	1.83	1.89
1.20	1.29	1.37	1.50	1.58	1.64	1.72	1.81	1.88
1.30	1.25	1.33	1.46	1.55	1.62	1.70	1.80	1.87
1.40	1.21	1.30	1.43	1.52	1.59	1.68	1.78	1.87
1.50	1.18	1.26	1.40	1.49	1.57	1.66	1.77	1.86
1.60	1.15	1.23	1.37	1.47	1.54	1.64	1.75	1.85
1.70	1.12	1.20	1.34	1.44	1.52	1.62	1.74	1.84
1.80	1.10	1.17	1.31	1.41	1.49	1.60	1.72	1.83
1.90	1.08	1.15	1.28	1.39	1.47	1.58	1.71	1.82
2.00	1.06	1.12	1.26	1.36	1.45	1.56	1.70	1.81
2.5	1.01	1.04	1.15	1.25	1.34	1.47	1.63	1.77
3.0	-	1.01	1.07	1.16	1.25	1.39	1.56	1.72
4.0	-	-	1.00	1.04	1.11	1.25	1.45	1.64
5.0	-	-	-	1.00	1.03	1.14	1.34	1.56
6.0	-	-	-	-	1.00	1.06	1.25	1.49
8.0	-	-	-	-	-	1.00	1.11	1.36
12.0	-	-	-	-	-	-	1.00	1.16
20.0	-	-	-	-	-	-	-	1.00

Table 18: The Effective Cost of Production when the cost of rework is ONE percent of the cost of scrap and the average is on the REWORK side of the target

C_{pk}	C_p values 0.10	0.20	0.30	0.40	0.50	0.60	0.70	0.80	0.90	1.00
1.00	-	-	-	-	-	-	-	-	-	1.06
0.90	-	-	-	-	-	-	-	-	1.07	1.03
0.80	-	-	-	-	-	-	-	1.09	1.04	1.02
0.70	-	-	-	-	-	-	1.11	1.05	1.03	1.01
0.60	-	-	-	-	-	1.14	1.07	1.03	1.02	1.01
0.50	-	-	-	-	1.19	1.09	1.04	1.02	1.01	1.01
0.40	-	-	-	1.25	1.12	1.05	1.02	1.01	1.01	1.01
0.30	-	-	1.34	1.16	1.07	1.03	1.01	1.01	1.01	1.01
0.20	-	1.48	1.22	1.10	1.04	1.02	1.01	1.01	1.01	1.01
0.10	1.69	1.31	1.13	1.06	1.03	1.01	1.01	1.01	1.01	1.01
0.00	1.43	1.19	1.08	1.03	1.02	1.01	1.01	1.01	1.01	1.01
–0.10	1.27	1.11	1.05	1.02	1.01	1.01	1.01	1.01	1.01	1.01
–0.20	1.16	1.07	1.03	1.01	1.01	1.01	1.01	1.01	1.01	1.01
–0.30	1.10	1.04	1.02	1.01	1.01	1.01	1.01	1.01	1.01	1.01
–0.40	1.06	1.02	1.01	1.01	1.01	1.01	1.01	1.01	1.01	1.01
–0.50	1.03	1.02	1.01	1.01	1.01	1.01	1.01	1.01	1.01	1.01

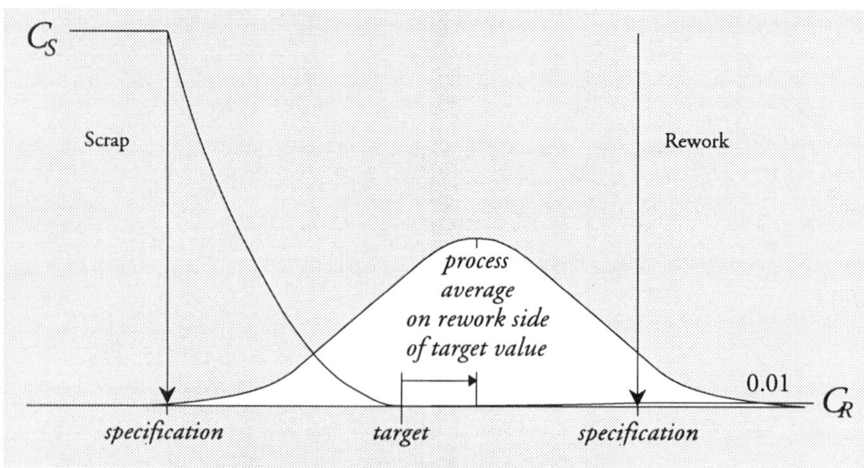

Table 18: Effective Cost of Production When Cost of Rework is 1% the Cost of Scrap and the Average is on the Rework Side of the Target Value

Table 18: The Effective Cost of Production when the cost of rework is ONE percent of the cost of scrap and the average is on the REWORK side of the target

	C_p values									
C_{pk}	1.10	1.20	1.30	1.40	1.50	1.60	1.70	1.80	1.90	2.00
2.00	-	-	-	-	-	-	-	-	-	1.01
1.90	-	-	-	-	-	-	-	-	1.02	1.01
1.80	-	-	-	-	-	-	-	1.02	1.01	1.01
1.70	-	-	-	-	-	-	1.02	1.01	1.01	1.00
1.60	-	-	-	-	-	1.02	1.01	1.01	1.00	1.00
1.50	-	-	-	-	1.02	1.01	1.01	1.00	1.00	1.00
1.40	-	-	-	1.03	1.02	1.01	1.00	1.00	1.00	1.00
1.30	-	-	1.03	1.02	1.01	1.00	1.00	1.00	1.00	1.00
1.20	-	1.04	1.02	1.01	1.01	1.00	1.00	1.00	1.00	1.00
1.10	1.05	1.02	1.01	1.01	1.00	1.00	1.00	1.00	1.00	1.00
1.00	1.03	1.01	1.01	1.00	1.00	1.00	1.00	1.00	1.00	1.00
0.90	1.02	1.01	1.00	1.00	1.00	1.00	1.00	1.00	1.00	1.00
0.80	1.01	1.01	1.00	1.00	1.00	1.00	1.00	1.00	1.00	1.00
0.70	1.01	1.00	1.00	1.00	1.00	1.00	1.00	1.00	1.00	1.00
0.60	1.00	1.00	1.00	1.00	1.00	1.00	1.00	1.00	1.00	1.01
0.50	1.00	1.00	1.00	1.00	1.00	1.01	1.01	1.01	1.01	1.01
0.40	1.00	1.00	1.01	1.01	1.01	1.01	1.01	1.01	1.01	1.01
0.30	1.01	1.01	1.01	1.01	1.01	1.01	1.01	1.01	1.01	1.01
0.20	1.01	1.01	1.01	1.01	1.01	1.01	1.01	1.01	1.01	1.01
0.10	1.01	1.01	1.01	1.01	1.01	1.01	1.01	1.01	1.01	1.01
0.00	1.01	1.01	1.01	1.01	1.01	1.01	1.01	1.01	1.01	1.01
–0.10	1.01	1.01	1.01	1.01	1.01	1.01	1.01	1.01	1.01	1.01
–0.20	1.01	1.01	1.01	1.01	1.01	1.01	1.01	1.01	1.01	1.01
–0.30	1.01	1.01	1.01	1.01	1.01	1.01	1.01	1.01	1.01	1.01
–0.40	1.01	1.01	1.01	1.01	1.01	1.01	1.01	1.01	1.01	1.01
–0.50	1.01	1.01	1.01	1.01	1.01	1.01	1.01	1.01	1.01	1.01

Blank spaces correspond to impossible combinations of C_p and C_{pk}.

When $C_{pk} = C_p$ the average is at the target.

Table 18: The Effective Cost of Production when the cost of rework is ONE percent of the cost of scrap and the average is on the REWORK side of the target

C_{pk}	C_p values 2.5	3.0	4.0	5.0	6.0	8.0	12.0	20.0
20.0	-	-	-	-	-	-	-	1.00
12.0	-	-	-	-	-	-	1.00	1.00
8.0	-	-	-	-	-	1.00	1.00	1.00
6.0	-	-	-	-	1.00	1.00	1.00	1.00
5.0	-	-	-	1.00	1.00	1.00	1.00	1.01
4.0	-	-	1.00	1.00	1.00	1.00	1.00	1.01
3.0	-	1.01	1.00	1.00	1.00	1.00	1.01	1.01
2.5	1.01	1.00	1.00	1.00	1.00	1.00	1.01	1.01
2.00	1.00	1.00	1.00	1.00	1.00	1.01	1.01	1.01
1.90	1.00	1.00	1.00	1.00	1.00	1.01	1.01	1.01
1.80	1.00	1.00	1.00	1.00	1.00	1.01	1.01	1.01
1.70	1.00	1.00	1.00	1.00	1.01	1.01	1.01	1.01
1.60	1.00	1.00	1.00	1.00	1.01	1.01	1.01	1.01
1.50	1.00	1.00	1.00	1.00	1.01	1.01	1.01	1.01
1.40	1.00	1.00	1.00	1.01	1.01	1.01	1.01	1.01
1.30	1.00	1.00	1.00	1.01	1.01	1.01	1.01	1.01
1.20	1.00	1.00	1.00	1.01	1.01	1.01	1.01	1.01
1.10	1.00	1.00	1.01	1.01	1.01	1.01	1.01	1.01
1.00	1.00	1.00	1.01	1.01	1.01	1.01	1.01	1.01
0.90	1.00	1.01	1.01	1.01	1.01	1.01	1.01	1.01
0.80	1.00	1.01	1.01	1.01	1.01	1.01	1.01	1.01
0.70	1.01	1.01	1.01	1.01	1.01	1.01	1.01	1.01
0.60	1.01	1.01	1.01	1.01	1.01	1.01	1.01	1.01
0.50	1.01	1.01	1.01	1.01	1.01	1.01	1.01	1.01
0.40	1.01	1.01	1.01	1.01	1.01	1.01	1.01	1.01
0.30	1.01	1.01	1.01	1.01	1.01	1.01	1.01	1.01
0.20	1.01	1.01	1.01	1.01	1.01	1.01	1.01	1.01
0.10	1.01	1.01	1.01	1.01	1.01	1.01	1.01	1.01
0.00	1.01	1.01	1.01	1.01	1.01	1.01	1.01	1.01
–0.10	1.01	1.01	1.01	1.01	1.01	1.01	1.01	1.01
–0.20	1.01	1.01	1.01	1.01	1.01	1.01	1.01	1.01
–0.30	1.01	1.01	1.01	1.01	1.01	1.01	1.01	1.01
–0.40	1.01	1.01	1.01	1.01	1.01	1.01	1.01	1.01
–0.50	1.01	1.01	1.01	1.01	1.01	1.01	1.01	1.01

Table 19: The Effective Cost of Production when the cost of rework is 0.1% of the cost of scrap and the average is on the SCRAP side of the target

C_{pk}	C_p values 0.10	0.20	0.30	0.40	0.50	0.60	0.70	0.80	0.90	1.00
–0.50	15.16	15.30	15.41	15.49	15.55	15.61	15.65	15.68	15.71	15.73
–0.40	8.86	8.98	9.09	9.17	9.24	9.29	9.33	9.37	9.40	9.42
–0.30	5.58	5.69	5.79	5.87	5.93	5.99	6.04	6.07	6.11	6.13
–0.20	3.77	3.87	3.96	4.04	4.10	4.16	4.21	4.25	4.28	4.31
–0.10	2.72	2.81	2.89	2.96	3.03	3.08	3.13	3.17	3.21	3.24
0.00	2.08	2.16	2.23	2.30	2.36	2.42	2.47	2.51	2.55	2.58
0.10	1.68	1.75	1.81	1.87	1.93	1.98	2.03	2.07	2.11	2.15
0.20	-	1.48	1.53	1.58	1.64	1.69	1.73	1.78	1.82	1.85
0.30	-	-	1.34	1.39	1.43	1.48	1.52	1.56	1.60	1.64
0.40	-	-	-	1.25	1.29	1.33	1.37	1.41	1.45	1.48
0.50	-	-	-	-	1.19	1.22	1.25	1.29	1.33	1.36
0.60	-	-	-	-	-	1.14	1.17	1.20	1.23	1.26
0.70	-	-	-	-	-	-	1.11	1.13	1.16	1.19
0.80	-	-	-	-	-	-	-	1.09	1.11	1.13
0.90	-	-	-	-	-	-	-	-	1.07	1.09
1.00	-	-	-	-	-	-	-	-	-	1.06

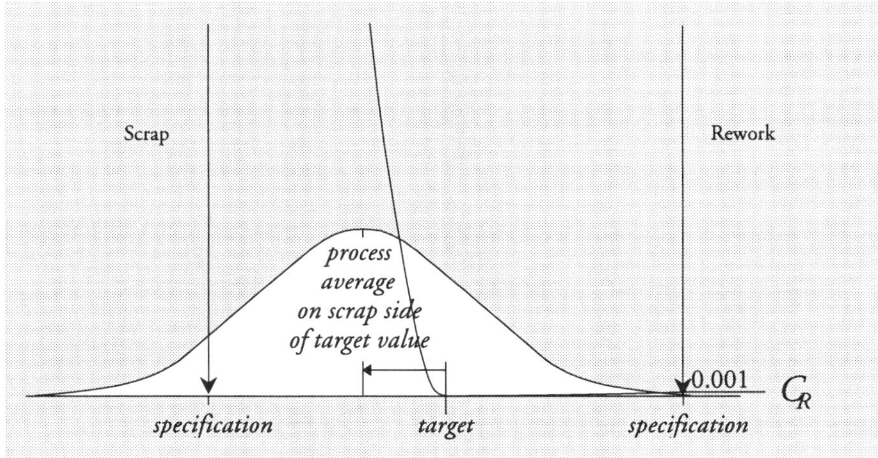

Table 19: Effective Cost of Production
When Cost of Rework is 0.1% the Cost of Scrap
and the Average is on the Scrap Side of the Target Value

Table 19: The Effective Cost of Production when the cost of rework is 0.1% of the cost of scrap and the average is on the SCRAP side of the target

C_{pk}	C_p values 1.10	1.20	1.30	1.40	1.50	1.60	1.70	1.80	1.90	2.00
–0.50	15.75	15.77	15.78	15.80	15.81	15.82	15.83	15.83	15.84	15.85
–0.40	9.44	9.46	9.48	9.49	9.51	9.52	9.53	9.53	9.54	9.55
–0.30	6.16	6.18	6.19	6.21	6.22	6.23	6.24	6.25	6.26	6.27
–0.20	4.33	4.36	4.38	4.39	4.41	4.42	4.43	4.44	4.45	4.46
–0.10	3.27	3.29	3.31	3.33	3.35	3.36	3.37	3.39	3.40	3.41
0.00	2.61	2.63	2.66	2.68	2.70	2.71	2.73	2.74	2.75	2.76
0.10	2.18	2.21	2.23	2.25	2.27	2.29	2.31	2.32	2.34	2.35
0.20	1.89	1.92	1.94	1.97	1.99	2.01	2.03	2.04	2.06	2.07
0.30	1.68	1.71	1.74	1.76	1.78	1.81	1.83	1.84	1.86	1.87
0.40	1.52	1.55	1.58	1.61	1.63	1.66	1.68	1.70	1.71	1.73
0.50	1.39	1.43	1.46	1.49	1.51	1.54	1.56	1.58	1.60	1.62
0.60	1.30	1.33	1.36	1.39	1.42	1.44	1.47	1.49	1.51	1.53
0.70	1.22	1.25	1.28	1.31	1.34	1.36	1.39	1.41	1.43	1.46
0.80	1.16	1.18	1.21	1.24	1.27	1.29	1.32	1.34	1.37	1.39
0.90	1.11	1.13	1.16	1.18	1.21	1.23	1.26	1.28	1.31	1.33
1.00	1.07	1.09	1.11	1.14	1.16	1.18	1.21	1.23	1.26	1.28
1.10	1.05	1.06	1.08	1.10	1.12	1.14	1.16	1.19	1.21	1.23
1.20	-	1.04	1.05	1.07	1.08	1.10	1.12	1.15	1.17	1.19
1.30	-	-	1.03	1.04	1.06	1.07	1.09	1.11	1.13	1.15
1.40	-	-	-	1.03	1.04	1.05	1.07	1.08	1.10	1.12
1.50	-	-	-	-	1.02	1.03	1.05	1.06	1.07	1.09
1.60	-	-	-	-	-	1.02	1.03	1.04	1.05	1.07
1.70	-	-	-	-	-	-	1.02	1.03	1.04	1.05
1.80	-	-	-	-	-	-	-	1.02	1.02	1.03
1.90	-	-	-	-	-	-	-	-	1.02	1.02
2.00	-	-	-	-	-	-	-	-	-	1.01

Blank spaces correspond to impossible combinations of C_p and C_{pk}.

When $C_{pk} = C_p$ the average is at the target.

Table 19: The Effective Cost of Production when the cost of rework is 0.1% of the cost of scrap and the average is on the SCRAP side of the target

C_{pk}	C_p values 2.5	3.0	4.0	5.0	6.0	8.0	12.0	20.0
–0.50	15.91	15.93	15.95	15.97	15.98	15.99	16.00	16.01
–0.40	9.59	9.61	9.64	9.65	9.66	9.68	9.69	9.70
–0.30	6.31	6.33	6.36	6.38	6.39	6.40	6.42	6.43
–0.20	4.50	4.52	4.56	4.57	4.59	4.60	4.62	4.63
–0.10	3.45	3.48	3.51	3.53	3.55	3.56	3.58	3.60
0.00	2.81	2.84	2.88	2.90	2.92	2.94	2.96	2.98
0.10	2.40	2.43	2.48	2.50	2.52	2.55	2.57	2.59
0.20	2.12	2.16	2.21	2.24	2.27	2.29	2.32	2.34
0.30	1.94	1.98	2.04	2.07	2.10	2.13	2.16	2.19
0.40	1.80	1.85	1.91	1.95	1.98	2.02	2.05	2.08
0.50	1.70	1.75	1.82	1.87	1.90	1.94	1.98	2.02
0.60	1.61	1.67	1.75	1.81	1.84	1.89	1.94	1.98
0.70	1.54	1.61	1.70	1.76	1.80	1.85	1.90	1.95
0.80	1.48	1.55	1.65	1.72	1.76	1.82	1.88	1.93
0.90	1.43	1.50	1.61	1.68	1.73	1.79	1.86	1.92
1.00	1.38	1.46	1.57	1.65	1.70	1.77	1.84	1.90
1.10	1.33	1.41	1.53	1.61	1.67	1.75	1.83	1.89
1.20	1.29	1.37	1.50	1.58	1.64	1.72	1.81	1.88
1.30	1.25	1.33	1.46	1.55	1.62	1.70	1.80	1.87
1.40	1.21	1.30	1.43	1.52	1.59	1.68	1.78	1.87
1.50	1.18	1.26	1.40	1.49	1.57	1.66	1.77	1.86
1.60	1.15	1.23	1.37	1.47	1.54	1.64	1.75	1.85
1.70	1.12	1.20	1.34	1.44	1.52	1.62	1.74	1.84
1.80	1.10	1.17	1.31	1.41	1.49	1.60	1.72	1.83
1.90	1.08	1.15	1.28	1.39	1.47	1.58	1.71	1.82
2.00	1.06	1.12	1.26	1.36	1.45	1.56	1.70	1.81
2.5	1.01	1.04	1.15	1.25	1.34	1.47	1.63	1.77
3.0	-	1.01	1.07	1.16	1.25	1.39	1.56	1.72
4.0	-	-	1.00	1.04	1.11	1.25	1.45	1.64
5.0	-	-	-	1.00	1.03	1.14	1.34	1.56
6.0	-	-	-	-	1.00	1.06	1.25	1.49
8.0	-	-	-	-	-	1.00	1.11	1.36
12.0	-	-	-	-	-	-	1.00	1.16
20.0	-	-	-	-	-	-	-	1.00

Table 20: The Effective Cost of Production when the cost of rework is 0.1% of the cost of scrap and the average is on the REWORK side of the target

C_{pk}	C_p values 0.10	0.20	0.30	0.40	0.50	0.60	0.70	0.80	0.90	1.00
1.00	-	-	-	-	-	-	-	-	-	1.06
0.90	-	-	-	-	-	-	-	-	1.07	1.03
0.80	-	-	-	-	-	-	-	1.09	1.04	1.02
0.70	-	-	-	-	-	-	1.11	1.05	1.02	1.01
0.60	-	-	-	-	-	1.14	1.07	1.03	1.01	1.01
0.50	-	-	-	-	1.19	1.209	1.04	1.02	1.01	1.00
0.40	-	-	-	1.25	1.12	1.05	1.02	1.01	1.00	1.00
0.30	-	-	1.34	1.16	1.07	1.03	1.01	1.00	1.00	1.00
0.20	-	1.48	1.21	1.09	1.04	1.01	1.01	1.00	1.00	1.00
0.10	1.68	1.30	1.13	1.05	1.02	1.01	1.00	1.00	1.00	1.00
0.00	1.43	1.18	1.07	1.03	1.01	1.00	1.00	1.00	1.00	1.00
–0.10	1.26	1.11	1.04	1.01	1.01	1.00	1.00	1.00	1.00	1.00
–0.20	1.16	1.06	1.02	1.01	1.00	1.00	1.00	1.00	1.00	1.00
–0.30	1.09	1.03	1.01	1.00	1.00	1.00	1.00	1.00	1.00	1.00
–0.40	1.05	1.02	1.01	1.00	1.00	1.00	1.00	1.00	1.00	1.00
–0.50	1.02	1.01	1.00	1.00	1.00	1.00	1.00	1.00	1.00	1.00

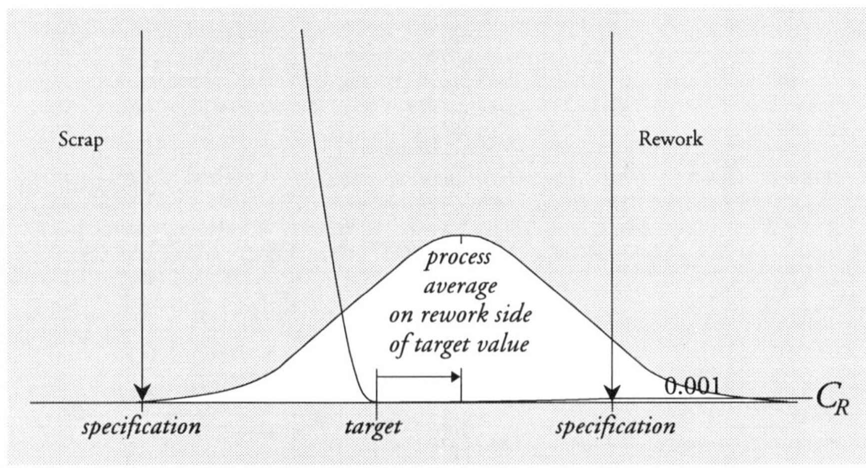

Table 20: Effective Cost of Production
When Cost of Rework is 0.1% the Cost of Scrap
and the Average is on the Rework Side of the Target Value

Table 20: The Effective Cost of Production when the cost of rework is 0.1% of the cost of scrap and the average is on the REWORK side of the target

C_{pk}	C_p values: 1.10	1.20	1.30	1.40	1.50	1.60	1.70	1.80	1.90	2.00
2.00	-	-	-	-	-	-	-	-	-	1.01
1.90	-	-	-	-	-	-	-	-	1.02	1.01
1.80	-	-	-	-	-	-	-	1.02	1.01	1.00
1.70	-	-	-	-	-	-	1.02	1.01	1.01	1.00
1.60	-	-	-	-	-	1.02	1.01	1.01	1.00	1.00
1.50	-	-	-	-	1.02	1.01	1.01	1.00	1.00	1.00
1.40	-	-	-	1.03	1.01	1.01	1.00	1.00	1.00	1.00
1.30	-	-	1.03	1.02	1.01	1.00	1.00	1.00	1.00	1.00
1.20	-	1.04	1.02	1.01	1.00	1.00	1.00	1.00	1.00	1.00
1.10	1.05	1.02	1.01	1.01	1.00	1.00	1.00	1.00	1.00	1.00
1.00	1.03	1.01	1.01	1.00	1.00	1.00	1.00	1.00	1.00	1.00
0.90	1.02	1.01	1.00	1.00	1.00	1.00	1.00	1.00	1.00	1.00
0.80	1.01	1.00	1.00	1.00	1.00	1.00	1.00	1.00	1.00	1.00
0.70	1.00	1.00	1.00	1.00	1.00	1.00	1.00	1.00	1.00	1.00
0.60	1.00	1.00	1.00	1.00	1.00	1.00	1.00	1.00	1.00	1.00
0.50	1.00	1.00	1.00	1.00	1.00	1.00	1.00	1.00	1.00	1.00
0.40	1.00	1.00	1.00	1.00	1.00	1.00	1.00	1.00	1.00	1.00
0.30	1.00	1.00	1.00	1.00	1.00	1.00	1.00	1.00	1.00	1.00
0.20	1.00	1.00	1.00	1.00	1.00	1.00	1.00	1.00	1.00	1.00
0.10	1.00	1.00	1.00	1.00	1.00	1.00	1.00	1.00	1.00	1.00
0.00	1.00	1.00	1.00	1.00	1.00	1.00	1.00	1.00	1.00	1.00
–0.10	1.00	1.00	1.00	1.00	1.00	1.00	1.00	1.00	1.00	1.00
–0.20	1.00	1.00	1.00	1.00	1.00	1.00	1.00	1.00	1.00	1.00
–0.30	1.00	1.00	1.00	1.00	1.00	1.00	1.00	1.00	1.00	1.00
–0.40	1.00	1.00	1.00	1.00	1.00	1.00	1.00	1.00	1.00	1.00
–0.50	1.00	1.00	1.00	1.00	1.00	1.00	1.00	1.00	1.00	1.00

Blank spaces correspond to impossible combinations of C_p and C_{pk}.

When $C_{pk} = C_p$ the average is at the target.

Table 20: The Effective Cost of Production when the cost of rework is 0.1% of the cost of scrap and the average is on the REWORK side of the target

C_{pk}	C_p values 2.5	3.0	4.0	5.0	6.0	8.0	12.0	20.0
20.0	-	-	-	-	-	-	-	1.00
12.0	-	-	-	-	-	-	1.00	1.00
8.0	-	-	-	-	-	1.00	1.00	1.00
6.0	-	-	-	-	1.00	1.00	1.00	1.00
5.0	-	-	-	1.00	1.00	1.00	1.00	1.00
4.0	-	-	1.00	1.00	1.00	1.00	1.00	1.00
3.0	-	1.01	1.00	1.00	1.00	1.00	1.00	1.00
2.5	1.01	1.00	1.00	1.00	1.00	1.00	1.00	1.00
2.00	1.00	1.00	1.00	1.00	1.00	1.00	1.00	1.00
1.90	1.00	1.00	1.00	1.00	1.00	1.00	1.00	1.00
1.80	1.00	1.00	1.00	1.00	1.00	1.00	1.00	1.00
1.70	1.00	1.00	1.00	1.00	1.00	1.00	1.00	1.00
1.60	1.00	1.00	1.00	1.00	1.00	1.00	1.00	1.00
1.50	1.00	1.00	1.00	1.00	1.00	1.00	1.00	1.00
1.40	1.00	1.00	1.00	1.00	1.00	1.00	1.00	1.00
1.30	1.00	1.00	1.00	1.00	1.00	1.00	1.00	1.00
1.20	1.00	1.00	1.00	1.00	1.00	1.00	1.00	1.00
1.10	1.00	1.00	1.00	1.00	1.00	1.00	1.00	1.00
1.00	1.00	1.00	1.00	1.00	1.00	1.00	1.00	1.00
0.90	1.00	1.00	1.00	1.00	1.00	1.00	1.00	1.00
0.80	1.00	1.00	1.00	1.00	1.00	1.00	1.00	1.00
0.70	1.00	1.00	1.00	1.00	1.00	1.00	1.00	1.00
0.60	1.00	1.00	1.00	1.00	1.00	1.00	1.00	1.00
0.50	1.00	1.00	1.00	1.00	1.00	1.00	1.00	1.00
0.40	1.00	1.00	1.00	1.00	1.00	1.00	1.00	1.00
0.30	1.00	1.00	1.00	1.00	1.00	1.00	1.00	1.00
0.20	1.00	1.00	1.00	1.00	1.00	1.00	1.00	1.00
0.10	1.00	1.00	1.00	1.00	1.00	1.00	1.00	1.00
0.00	1.00	1.00	1.00	1.00	1.00	1.00	1.00	1.00
–0.10	1.00	1.00	1.00	1.00	1.00	1.00	1.00	1.00
–0.20	1.00	1.00	1.00	1.00	1.00	1.00	1.00	1.00
–0.30	1.00	1.00	1.00	1.00	1.00	1.00	1.00	1.00
–0.40	1.00	1.00	1.00	1.00	1.00	1.00	1.00	1.00
–0.50	1.00	1.00	1.00	1.00	1.00	1.00	1.00	1.00

Tables 21 to 29
The Effective Cost of Production for One-Sided Product Specifications

Table 21: The Effective Cost of Production for a one-sided product specification without a stated target value

	Scrap	*Rework (cost of rework / nominal cost)*						
C_{pk}		***1.00***	***0.80***	***0.67***	***0.50***	***0.33***	***0.20***	***0.10***
–0.50	15.73	1.983	1.786	1.655	1.491	1.328	1.197	1.098
–0.40	9.42	1.968	1.774	1.645	1.484	1.323	1.194	1.097
–0.30	6.13	1.943	1.755	1.629	1.472	1.314	1.189	1.094
–0.20	4.31	1.907	1.725	1.604	1.453	1.302	1.181	1.091
–0.10	3.24	1.856	1.684	1.570	1.428	1.285	1.171	1.086
0.00	2.58	1.789	1.631	1.526	1.395	1.263	1.158	1.079
0.10	2.15	1.709	1.568	1.473	1.355	1.236	1.142	1.071
0.20	1.853	1.619	1.495	1.413	1.310	1.206	1.124	1.062
0.30	1.641	1.523	1.419	1.349	1.262	1.174	1.105	1.052
0.40	1.483	1.427	1.342	1.285	1.214	1.142	1.085	1.043
0.50	1.361	1.336	1.269	1.224	1.168	1.112	1.067	1.034
0.60	1.265	1.255	1.204	1.170	1.128	1.085	1.051	1.026
0.70	1.189	1.186	1.149	1.124	1.093	1.062	1.037	1.019
0.80	1.131	1.130	1.104	1.087	1.065	1.043	1.026	1.013
0.90	1.087	1.087	1.069	1.058	1.043	1.029	1.017	1.009
1.00	1.055	1.055	1.044	1.037	1.028	1.018	1.011	1.006
1.10	1.033	1.033	1.027	1.022	1.017	1.011	1.007	1.003
1.20	1.019	1.019	1.015	1.013	1.010	1.006	1.004	1.002
1.30	1.010	1.010	1.008	1.007	1.005	1.003	1.002	1.001
1.40	1.005	1.005	1.004	1.004	1.003	1.002	1.001	1.001
1.50	1.003	1.003	1.002	1.002	1.001	1.001	1.001	1.000
1.60	1.001	1.001	1.001	1.001	1.001	1.000	1.000	1.000
1.70	1.000	1.000	1.000	1.000	1.000	1.000	1.000	1.000

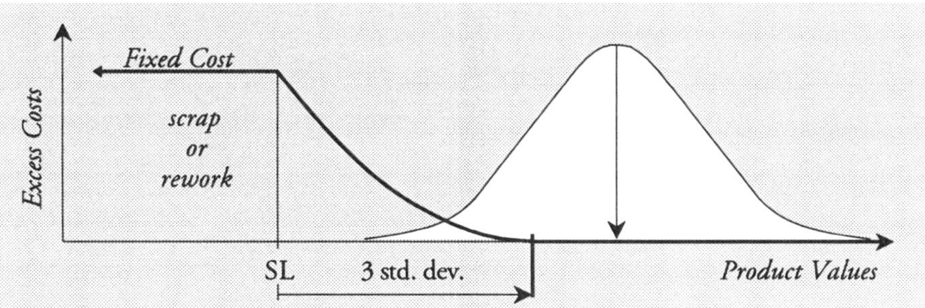

Table 21: One-Sided Product Specification Without a Stated Target Value

Table 22: The Effective Cost of Production for a one-sided specification with stated target when nonconforming product is scrapped

C_{pk}	Tolerance in Standard Deviation Units										
	0.0	***0.3***	***0.6***	***0.9***	***1.2***	***1.5***	***1.8***	***2.1***	***2.4***	***2.7***	***3.0***
–0.50	14.99	15.16	15.30	15.41	15.49	15.55	15.61	15.65	15.68	15.71	15.73
–0.40	8.70	8.86	8.98	9.17	9.17	9.24	9.29	9.33	9.37	9.40	9.42
–0.30	5.44	5.58	5.69	5.79	5.87	5.93	5.99	6.04	6.07	6.11	6.13
–0.20	3.65	3.77	3.87	3.96	4.04	4.10	4.16	4.21	4.25	4.28	4.31
–0.10	2.62	2.72	2.81	2.89	2.96	3.03	3.08	3.13	3.17	3.21	3.24
0.00	2.00	2.08	2.16	2.23	2.30	2.36	2.42	2.47	2.51	2.55	2.58
0.10	1.62	1.68	1.75	1.81	1.87	1.93	1.98	2.03	2.07	2.11	2.15
0.20	1.38	1.43	1.48	1.53	1.58	1.64	1.69	1.73	1.78	1.82	1.85
0.30	1.23	1.26	1.30	1.34	1.39	1.43	1.48	1.52	1.56	1.60	1.64
0.40	1.13	1.15	1.18	1.21	1.25	1.29	1.33	1.37	1.41	1.45	1.48
0.50	1.07	1.09	1.11	1.13	1.16	1.19	1.22	1.25	1.29	1.33	1.36
0.60	1.04	1.05	1.06	1.07	1.09	1.11	1.14	1.17	1.20	1.23	1.26
0.70	1.02	1.02	1.03	1.04	1.05	1.07	1.09	1.11	1.13	1.16	1.19
0.80	1.01	1.01	1.01	1.02	1.03	1.04	1.05	1.07	1.08	1.11	1.13
0.90	1.00	1.00	1.01	1.01	1.01	1.02	1.03	1.04	1.05	1.07	1.09
1.00	1.00	1.00	1.00	1.00	1.01	1.01	1.01	1.02	1.03	1.04	1.06
1.10	1.00	1.00	1.00	1.00	1.00	1.00	1.01	1.01	1.02	1.02	1.03
1.20	1.00	1.00	1.00	1.00	1.00	1.00	1.00	1.01	1.01	1.01	1.02
1.30	1.00	1.00	1.00	1.00	1.00	1.00	1.00	1.00	1.00	1.01	1.01
1.40	1.00	1.00	1.00	1.00	1.00	1.00	1.00	1.00	1.00	1.00	1.01
1.50	1.00	1.00	1.00	1.00	1.00	1.00	1.00	1.00	1.00	1.00	1.00

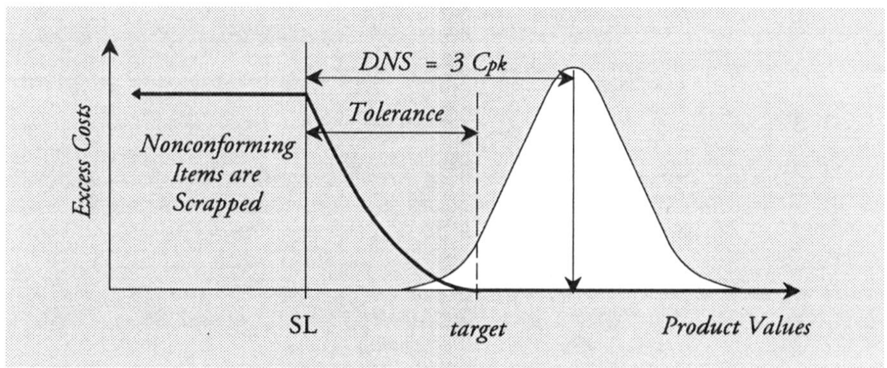

Table 22: One-Sided Specification With a Stated Target Value

Table 22: The Effective Cost of Production for a one-sided specification with stated target when nonconforming product is scrapped

Tolerance in Standard Deviation Units

C_{pk}	3.3	3.6	3.9	4.2	4.5	4.8	5.1	5.4	5.7	6.0
–0.50	15.75	15.77	15.78	15.80	15.81	15.82	15.83	15.83	15.84	15.85
–0.40	9.44	9.46	9.48	9.49	9.51	9.52	9.53	9.53	9.54	9.55
–0.30	6.16	6.18	6.19	6.21	6.22	6.23	6.24	6.25	6.26	6.27
–0.20	4.33	4.36	4.38	4.39	4.41	4.42	4.43	4.44	4.45	4.46
–0.10	3.27	3.29	3.31	3.33	3.35	3.36	3.37	3.39	3.40	3.41
0.00	2.61	2.63	2.66	2.68	2.70	2.71	2.73	2.74	2.75	2.76
0.10	2.18	2.21	2.23	2.25	2.27	2.29	2.31	2.32	2.34	2.35
0.20	1.89	1.92	1.94	1.97	1.99	2.01	2.03	2.04	2.06	2.07
0.30	1.68	1.71	1.74	1.76	1.78	1.81	1.83	1.84	1.86	1.87
0.40	1.52	1.55	1.58	1.61	1.63	1.66	1.68	1.70	1.71	1.73
0.50	1.39	1.43	1.46	1.49	1.51	1.54	1.56	1.58	1.60	1.62
0.60	1.30	1.33	1.36	1.39	1.42	1.44	1.47	1.49	1.51	1.53
0.70	1.22	1.25	1.28	1.31	1.34	1.36	1.39	1.41	1.43	1.46
0.80	1.16	1.18	1.21	1.24	1.27	1.29	1.32	1.34	1.37	1.39
0.90	1.11	1.13	1.16	1.18	1.21	1.23	1.26	1.28	1.31	1.33
1.00	1.07	1.09	1.11	1.14	1.16	1.18	1.21	1.23	1.26	1.28
1.10	1.05	1.06	1.08	1.10	1.12	1.14	1.16	1.19	1.21	1.23
1.20	1.03	1.04	1.05	1.07	1.08	1.10	1.12	1.15	1.17	1.19
1.30	1.02	1.02	1.03	1.04	1.06	1.07	1.09	1.11	1.13	1.15
1.40	1.01	1.01	1.02	1.03	1.04	1.05	1.07	1.08	1.10	1.12
1.50	1.00	1.01	1.01	1.02	1.02	1.03	1.05	1.06	1.07	1.09
1.60	1.00	1.00	1.01	1.01	1.01	1.02	1.03	1.04	1.05	1.07
1.70	1.00	1.00	1.00	1.01	1.01	1.01	1.02	1.03	1.04	1.05
1.80	1.00	1.00	1.00	1.00	1.00	1.01	1.01	1.02	1.02	1.03
1.90	1.00	1.00	1.00	1.00	1.00	1.00	1.01	1.01	1.02	1.02
2.00	1.00	1.00	1.00	1.00	1.00	1.00	1.00	1.01	1.01	1.01

Table 23: The Effective Cost of Production for a one-sided specification with stated target when nonconforming product is reworked and the cost of rework is equal to the nominal cost of production

C_{pk}	Tolerance in Standard Deviation Units 0.0	0.3	0.6	0.9	1.2	1.5	1.8	2.1	2.4	2.7	3.0
–0.50	1.93	1.94	1.95	1.96	1.97	1.97	1.97	1.98	1.98	1.98	1.98
–0.40	1.89	1.90	1.92	1.93	1.94	1.95	1.95	1.96	1.96	1.96	1.97
–0.30	1.82	1.84	1.86	1.88	1.89	1.91	1.92	1.93	1.93	1.94	1.94
–0.20	1.73	1.76	1.79	1.81	1.83	1.85	1.87	1.88	1.89	1.90	1.91
–0.10	1.62	1.66	1.69	1.72	1.75	1.77	1.80	1.81	1.83	1.84	1.86
0.00	1.50	1.54	1.58	1.62	1.65	1.68	1.71	1.73	1.75	1.77	1.79
0.10	1.38	1.42	1.46	1.50	1.54	1.57	1.61	1.64	1.66	1.69	1.71
0.20	1.27	1.31	1.35	1.38	1.42	1.46	1.50	1.53	1.56	1.59	1.62
0.30	1.18	1.21	1.24	1.28	1.31	1.35	1.39	1.43	1.46	1.49	1.52
0.40	1.12	1.14	1.16	1.19	1.22	1.25	1.29	1.32	1.36	1.39	1.43
0.50	1.07	1.08	1.10	1.12	1.15	1.17	1.20	1.24	1.27	1.30	1.34
0.60	1.04	1.05	1.06	1.07	1.09	1.11	1.14	1.16	1.19	1.22	1.26
0.70	1.02	1.02	1.03	1.04	1.05	1.07	1.08	1.11	1.13	1.16	1.19
0.80	1.01	1.01	1.01	1.02	1.03	1.04	1.05	1.07	1.08	1.11	1.13
0.90	1.00	1.00	1.01	1.01	1.01	1.02	1.03	1.04	1.05	1.07	1.09
1.00	1.00	1.00	1.00	1.00	1.01	1.01	1.01	1.02	1.03	1.04	1.06
1.10	1.00	1.00	1.00	1.00	1.00	1.00	1.01	1.01	1.02	1.02	1.03
1.20	1.00	1.00	1.00	1.00	1.00	1.00	1.00	1.01	1.01	1.01	1.02
1.30	1.00	1.00	1.00	1.00	1.00	1.00	1.00	1.00	1.00	1.01	1.01
1.40	1.00	1.00	1.00	1.00	1.00	1.00	1.00	1.00	1.00	1.00	1.01
1.50	1.00	1.00	1.00	1.00	1.00	1.00	1.00	1.00	1.00	1.00	1.00

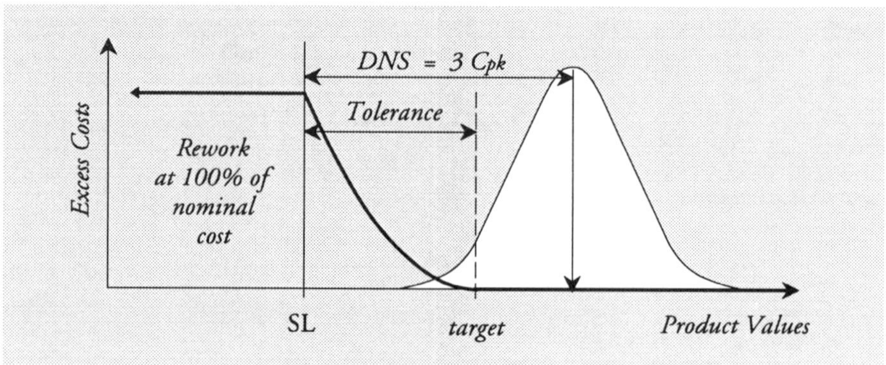

Table 23: One-Sided Specification With a Stated Target Value

Table 23: The Effective Cost of Production for a one-sided specification with stated target when nonconforming product is reworked and the cost of rework is equal to the nominal cost of production

Tolerance in Standard Deviation Units

C_{pk}	3.3	3.6	3.9	4.2	4.5	4.8	5.1	5.4	5.7	6.0
–0.50	1.98	1.99	1.99	1.99	1.99	1.99	1.99	1.99	1.99	1.99
–0.40	1.97	1.97	1.97	1.98	1.98	1.98	1.98	1.98	1.98	1.98
–0.30	1.95	1.95	1.95	1.96	1.96	1.96	1.96	1.97	1.97	1.97
–0.20	1.91	1.92	1.92	1.93	1.93	1.94	1.94	1.94	1.95	1.95
–0.10	1.87	1.87	1.88	1.89	1.90	1.90	1.91	1.91	1.91	1.92
0.00	1.80	1.82	1.83	1.84	1.85	1.85	1.86	1.87	1.87	1.88
0.10	1.73	1.75	1.76	1.77	1.79	1.80	1.81	1.82	1.82	1.83
0.20	1.64	1.66	1.68	1.70	1.72	1.73	1.74	1.76	1.77	1.78
0.30	1.55	1.58	1.60	1.62	1.64	1.66	1.67	1.69	1.70	1.71
0.40	1.46	1.49	1.51	1.54	1.56	1.58	1.60	1.62	1.63	1.65
0.50	1.37	1.40	1.43	1.45	1.48	1.50	1.52	1.54	1.56	1.58
0.60	1.29	1.32	1.35	1.38	1.40	1.43	1.45	1.47	1.49	1.51
0.70	1.22	1.25	1.27	1.30	1.33	1.36	1.38	1.41	1.43	1.45
0.80	1.16	1.18	1.21	1.24	1.27	1.29	1.32	1.34	1.36	1.39
0.90	1.11	1.13	1.16	1.18	1.21	1.23	1.26	1.28	1.31	1.33
1.00	1.07	1.09	1.11	1.14	1.16	1.18	1.21	1.23	1.25	1.28
1.10	1.05	1.06	1.08	1.10	1.12	1.14	1.16	1.19	1.21	1.23
1.20	1.03	1.04	1.05	1.07	1.08	1.10	1.12	1.15	1.17	1.19
1.30	1.02	1.02	1.03	1.04	1.06	1.07	1.09	1.11	1.13	1.15
1.40	1.01	1.01	1.02	1.03	1.04	1.05	1.07	1.08	1.10	1.12
1.50	1.00	1.01	1.01	1.02	1.02	1.03	1.05	1.06	1.07	1.09
1.60	1.00	1.00	1.01	1.01	1.01	1.02	1.03	1.04	1.05	1.07
1.70	1.00	1.00	1.00	1.01	1.01	1.01	1.02	1.03	1.04	1.05
1.80	1.00	1.00	1.00	1.00	1.00	1.01	1.01	1.02	1.02	1.03
1.90	1.00	1.00	1.00	1.00	1.00	1.00	1.01	1.01	1.02	1.02
2.00	1.00	1.00	1.00	1.00	1.00	1.00	1.00	1.01	1.01	1.01

Table 24: The Effective Cost of Production for a one-sided specification with stated target when nonconforming product is reworked and the cost of rework is 0.80 of the nominal cost of production

	Tolerance in Standard Deviation Units										
C_{pk}	0.0	0.3	0.6	0.9	1.2	1.5	1.8	2.1	2.4	2.7	3.0
–0.50	1.75	1.76	1.76	1.77	1.77	1.78	1.78	1.78	1.78	1.79	1.79
–0.40	1.71	1.72	1.73	1.74	1.75	1.76	1.76	1.77	1.77	1.77	1.77
–0.30	1.65	1.67	1.69	1.70	1.72	1.73	1.73	1.74	1.75	1.75	1.75
–0.20	1.58	1.61	1.63	1.65	1.67	1.68	1.69	1.70	1.71	1.72	1.73
–0.10	1.49	1.52	1.55	1.58	1.60	1.62	1.64	1.65	1.66	1.67	1.68
0.00	1.40	1.43	1.46	1.49	1.52	1.54	1.57	1.59	1.60	1.62	1.63
0.10	1.31	1.34	1.37	1.40	1.43	1.46	1.49	1.51	1.53	1.55	1.57
0.20	1.22	1.25	1.28	1.31	1.34	1.37	1.40	1.43	1.45	1.47	1.50
0.30	1.15	1.17	1.20	1.22	1.25	1.28	1.31	1.34	1.37	1.39	1.42
0.40	1.09	1.11	1.13	1.15	1.18	1.20	1.23	1.26	1.29	1.32	1.34
0.50	1.05	1.07	1.08	1.10	1.12	1.14	1.16	1.19	1.22	1.24	1.27
0.60	1.03	1.04	1.05	1.06	1.07	1.09	1.11	1.13	1.15	1.18	1.20
0.70	1.01	1.02	1.02	1.03	1.04	1.05	1.07	1.08	1.10	1.13	1.15
0.80	1.01	1.01	1.01	1.02	1.02	1.03	1.04	1.05	1.07	1.08	1.10
0.90	1.00	1.00	1.01	1.01	1.01	1.02	1.02	1.03	1.04	1.05	1.07
1.00	1.00	1.00	1.00	1.00	1.01	1.01	1.01	1.02	1.02	1.03	1.04
1.10	1.00	1.00	1.00	1.00	1.00	1.00	1.01	1.01	1.01	1.02	1.03
1.20	1.00	1.00	1.00	1.00	1.00	1.00	1.00	1.00	1.01	1.01	1.02
1.30	1.00	1.00	1.00	1.00	1.00	1.00	1.00	1.00	1.00	1.01	1.01
1.40	1.00	1.00	1.00	1.00	1.00	1.00	1.00	1.00	1.00	1.00	1.00
1.50	1.00	1.00	1.00	1.00	1.00	1.00	1.00	1.00	1.00	1.00	1.00

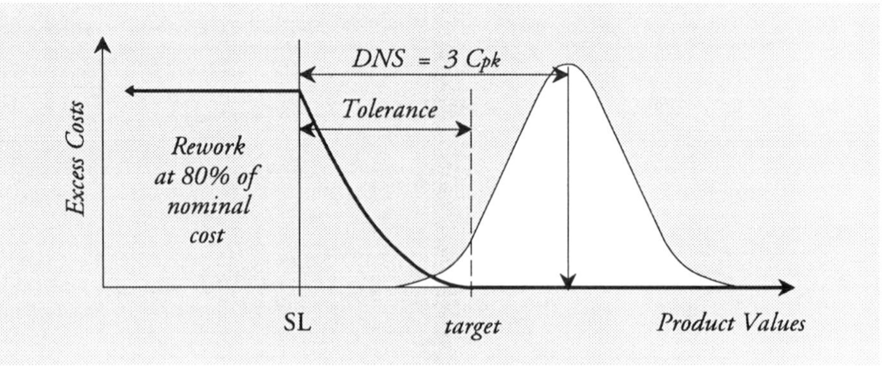

Table 24: One-Sided Specification With a Stated Target Value

Table 24: The Effective Cost of Production for a one-sided specification with stated target when nonconforming product is reworked and the cost of rework is 0.80 of the nominal cost of production

	Tolerance in Standard Deviation Units									
C_{pk}	**3.3**	**3.6**	**3.9**	**4.2**	**4.5**	**4.8**	**5.1**	**5.4**	**5.7**	**6.0**
–0.50	1.79	1.79	1.79	1.79	1.79	1.79	1.79	1.79	1.79	1.79
–0.40	1.78	1.78	1.78	1.78	1.78	1.78	1.78	1.78	1.79	1.79
–0.30	1.76	1.76	1.76	1.77	1.77	1.77	1.77	1.77	1.77	1.77
–0.20	1.73	1.74	1.74	1.74	1.75	1.75	1.75	1.75	1.76	1.76
–0.10	1.69	1.70	1.71	1.71	1.72	1.72	1.73	1.73	1.73	1.73
0.00	1.64	1.65	1.66	1.67	1.68	1.68	1.69	1.69	1.70	1.70
0.10	1.58	1.60	1.61	1.62	1.63	1.64	1.65	1.65	1.66	1.67
0.20	1.51	1.53	1.55	1.56	1.57	1.58	1.59	1.60	1.61	1.62
0.30	1.44	1.46	1.48	1.50	1.51	1.53	1.54	1.55	1.56	1.57
0.40	1.37	1.39	1.41	1.43	1.45	1.46	1.48	1.49	1.51	1.52
0.50	1.29	1.32	1.34	1.36	1.38	1.40	1.42	1.44	1.45	1.46
0.60	1.23	1.25	1.28	1.30	1.32	1.34	1.36	1.38	1.39	1.41
0.70	1.17	1.20	1.22	1.24	1.26	1.29	1.31	1.32	1.34	1.36
0.80	1.12	1.15	1.17	1.19	1.21	1.23	1.25	1.27	1.29	1.31
0.90	1.09	1.11	1.13	1.15	1.17	1.19	1.21	1.23	1.25	1.26
1.00	1.06	1.07	1.09	1.11	1.13	1.15	1.17	1.19	1.20	1.22
1.10	1.04	1.05	1.06	1.08	1.09	1.11	1.13	1.15	1.17	1.18
1.20	1.02	1.03	1.04	1.05	1.07	1.08	1.10	1.12	1.13	1.15
1.30	1.01	1.02	1.03	1.04	1.05	1.06	1.07	1.09	1.10	1.12
1.40	1.01	1.01	1.02	1.02	1.03	1.04	1.05	1.07	1.08	1.09
1.50	1.00	1.01	1.01	1.01	1.02	1.03	1.04	1.05	1.06	1.07
1.60	1.00	1.00	1.00	1.01	1.01	1.02	1.02	1.03	1.04	1.05
1.70	1.00	1.00	1.00	1.00	1.01	1.01	1.02	1.02	1.03	1.04
1.80	1.00	1.00	1.00	1.00	1.00	1.01	1.01	1.01	1.02	1.03
1.90	1.00	1.00	1.00	1.00	1.00	1.00	1.01	1.01	1.01	1.02
2.00	1.00	1.00	1.00	1.00	1.00	1.00	1.00	1.00	1.01	1.01

Table 25: The Effective Cost of Production for a one-sided specification with stated target when nonconforming product is reworked and the cost of rework is 0.67 of the nominal cost of production

	Tolerance in Standard Deviation Units										
C_{pk}	***0.0***	***0.3***	***0.6***	***0.9***	***1.2***	***1.5***	***1.8***	***2.1***	***2.4***	***2.7***	***3.0***
–0.50	1.62	1.63	1.64	1.64	1.64	1.65	1.65	1.65	1.65	1.65	1.66
–0.40	1.59	1.60	1.61	1.62	1.63	1.63	1.63	1.64	1.64	1.64	1.65
–0.30	1.54	1.56	1.57	1.59	1.30	1.60	1.61	1.62	1.62	1.63	1.63
–0.20	1.48	1.51	1.52	1.54	1.55	1.57	1.58	1.59	1.59	1.60	1.60
–0.10	1.41	1.44	1.46	1.48	1.50	1.52	1.53	1.54	1.55	1.56	1.57
0.00	1.33	1.36	1.39	1.41	1.43	1.45	1.47	1.49	1.50	1.52	1.53
0.10	1.25	1.28	1.31	1.33	1.36	1.38	1.40	1.42	1.44	1.46	1.47
0.20	1.18	1.21	1.23	1.26	1.28	1.31	1.33	1.35	1.38	1.40	1.41
0.30	1.12	1.14	1.16	1.19	1.21	1.23	1.26	1.28	1.31	1.33	1.35
0.40	1.08	1.09	1.11	1.13	1.15	1.17	1.19	1.22	1.24	1.26	1.28
0.50	1.04	1.05	1.07	1.08	1.10	1.12	1.14	1.16	1.18	1.20	1.22
0.60	1.02	1.03	1.04	1.05	1.06	1.07	1.09	1.11	1.13	1.15	1.17
0.70	1.01	1.02	1.02	1.03	1.03	1.04	1.06	1.07	1.09	1.10	1.12
0.80	1.00	1.01	1.01	1.01	1.02	1.02	1.03	1.04	1.06	1.07	1.09
0.90	1.00	1.00	1.00	1.01	1.01	1.01	1.02	1.03	1.03	1.05	1.06
1.00	1.00	1.00	1.00	1.00	1.00	1.01	1.01	1.01	1.02	1.03	1.04
1.10	1.00	1.00	1.00	1.00	1.00	1.00	1.00	1.01	1.01	1.02	1.02
1.20	1.00	1.00	1.00	1.00	1.00	1.00	1.00	1.00	1.01	1.01	1.01
1.30	1.00	1.00	1.00	1.00	1.00	1.00	1.00	1.00	1.00	1.00	1.01
1.40	1.00	1.00	1.00	1.00	1.00	1.00	1.00	1.00	1.00	1.00	1.00
1.50	1.00	1.00	1.00	1.00	1.00	1.00	1.00	1.00	1.00	1.00	1.00

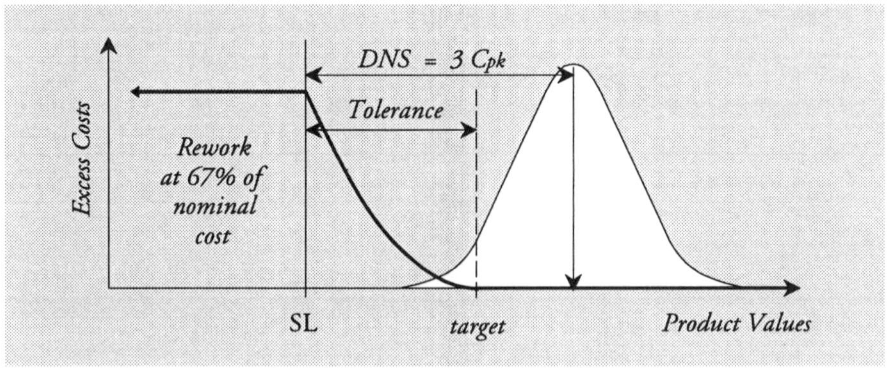

Table 25: One-Sided Specification With a Stated Target Value

Table 25: The Effective Cost of Production for a one-sided specification with stated target when nonconforming product is reworked and the cost of rework is 0.67 of the nominal cost of production

	Tolerance in Standard Deviation Units									
C_{pk}	**3.3**	**3.6**	**3.9**	**4.2**	**4.5**	**4.8**	**5.1**	**5.4**	**5.7**	**6.0**
–0.50	1.66	1.66	1.66	1.66	1.66	1.66	1.66	1.66	1.66	1.66
–0.40	1.65	1.65	1.65	1.65	1.65	1.65	1.65	1.65	1.65	1.65
–0.30	1.63	1.63	1.64	1.64	1.64	1.64	1.64	1.64	1.64	1.65
–0.20	1.61	1.61	1.62	1.62	1.62	1.62	1.63	1.63	1.63	1.63
–0.10	1.58	1.58	1.59	1.59	1.60	1.60	1.60	1.61	1.61	1.61
0.00	1.54	1.54	1.55	1.56	1.56	1.57	1.57	1.58	1.58	1.59
0.10	1.49	1.50	1.51	1.52	1.52	1.53	1.54	1.54	1.55	1.55
0.20	1.43	1.44	1.46	1.47	1.48	1.49	1.50	1.50	1.51	1.52
0.30	1.37	1.38	1.40	1.41	1.43	1.44	1.45	1.46	1.47	1.48
0.40	1.31	1.32	1.34	1.36	1.37	1.39	1.40	1.41	1.42	1.43
0.50	1.25	1.27	1.29	1.30	1.32	1.34	1.35	1.36	1.38	1.39
0.60	1.19	1.21	1.23	1.25	1.27	1.29	1.30	1.32	1.33	1.34
0.70	1.14	1.16	1.18	1.20	1.22	1.24	1.25	1.27	1.28	1.30
0.80	1.10	1.12	1.14	1.16	1.18	1.19	1.21	1.23	1.24	1.26
0.90	1.07	1.09	1.10	1.12	1.14	1.16	1.17	1.19	1.20	1.22
1.00	1.05	1.06	1.08	1.09	1.11	1.12	1.14	1.15	1.17	1.19
1.10	1.03	1.04	1.05	1.06	1.08	1.09	1.11	1.12	1.14	1.15
1.20	1.02	1.03	1.03	1.04	1.06	1.07	1.08	1.10	1.11	1.13
1.30	1.01	1.02	1.02	1.03	1.04	1.05	1.06	1.07	1.09	1.10
1.40	1.01	1.01	1.01	1.02	1.03	1.03	1.04	1.05	1.07	1.08
1.50	1.00	1.00	1.01	1.01	1.02	1.02	1.03	1.04	1.05	1.06
1.60	1.00	1.00	1.00	1.01	1.01	1.01	1.02	1.03	1.04	1.04
1.70	1.00	1.00	1.00	1.00	1.01	1.01	1.01	1.02	1.02	1.03
1.80	1.00	1.00	1.00	1.00	1.00	1.01	1.01	1.01	1.02	1.02
1.90	1.00	1.00	1.00	1.00	1.00	1.00	1.00	1.01	1.01	1.01
2.00	1.00	1.00	1.00	1.00	1.00	1.00	1.00	1.00	1.01	1.01

Table 26: The Effective Cost of Production for a one-sided specification with stated target when nonconforming product is reworked and the cost of rework is 0.50 of the nominal cost of production

	Tolerance in Standard Deviation Units										
C_{pk}	*0.0*	*0.3*	*0.6*	*0.9*	*1.2*	*1.5*	*1.8*	*2.1*	*2.4*	*2.7*	*3.0*
–0.50	1.47	1.47	1.48	1.48	1.48	1.49	1.49	1.49	1.49	1.49	1.49
–0.40	1.44	1.45	1.46	1.46	1.47	1.47	1.48	1.48	1.48	1.48	1.48
–0.30	1.41	1.42	1.43	1.44	1.45	1.45	1.46	1.46	1.47	1.47	1.47
–0.20	1.36	1.38	1.39	1.41	1.42	1.43	1.43	1.44	1.44	1.45	1.45
–0.10	1.31	1.33	1.34	1.36	1.37	1.39	1.40	1.41	1.41	1.42	1.43
0.00	1.25	1.27	1.29	1.31	1.32	1.34	1.35	1.37	1.38	1.39	1.39
0.10	1.19	1.21	1.23	1.25	1.27	1.29	1.30	1.32	1.33	1.34	1.35
0.20	1.14	1.15	1.17	1.19	1.21	1.23	1.25	1.27	1.28	1.30	1.31
0.30	1.09	1.11	1.12	1.14	1.16	1.18	1.19	1.21	1.23	1.25	1.26
0.40	1.06	1.07	1.08	1.09	1.11	1.13	1.14	1.16	1.18	1.20	1.21
0.50	1.03	1.04	1.05	1.06	1.07	1.09	1.10	1.12	1.13	1.15	1.17
0.60	1.02	1.02	1.03	1.04	1.04	1.06	1.07	1.08	1.10	1.11	1.13
0.70	1.01	1.01	1.02	1.02	1.03	1.03	1.04	1.05	1.07	1.08	1.09
0.80	1.00	1.01	1.01	1.01	1.01	1.02	1.02	1.03	1.04	1.05	1.06
0.90	1.00	1.00	1.00	1.00	1.01	1.01	1.01	1.02	1.03	1.03	1.04
1.00	1.00	1.00	1.00	1.00	1.00	1.00	1.01	1.01	1.01	1.02	1.03
1.10	1.00	1.00	1.00	1.00	1.00	1.00	1.00	1.01	1.01	1.01	1.02
1.20	1.00	1.00	1.00	1.00	1.00	1.00	1.00	1.00	1.00	1.01	1.01
1.30	1.00	1.00	1.00	1.00	1.00	1.00	1.00	1.00	1.00	1.00	1.01
1.40	1.00	1.00	1.00	1.00	1.00	1.00	1.00	1.00	1.00	1.00	1.00

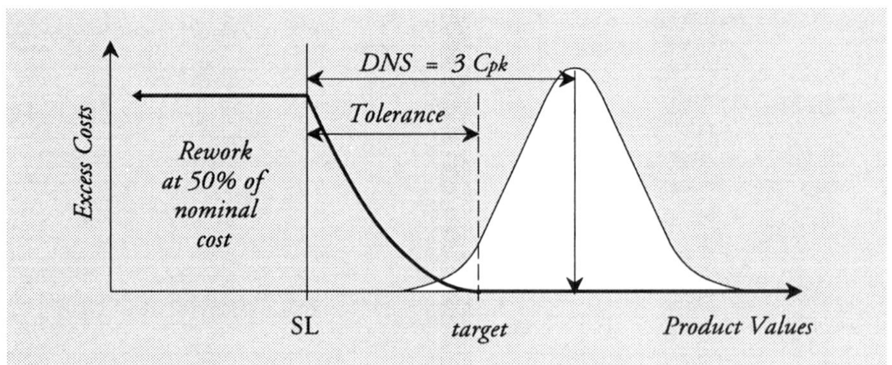

Table 26: One-Sided Specification With a Stated Target Value

Table 26: The Effective Cost of Production for a one-sided specification with stated target when nonconforming product is reworked and the cost of rework is 0.50 of the nominal cost of production

	Tolerance in Standard Deviation Units									
C_{pk}	***3.3***	***3.6***	***3.9***	***4.2***	***4.5***	***4.8***	***5.1***	***5.4***	***5.7***	***6.0***
–0.50	1.49	1.49	1.49	1.49	1.49	1.49	1.49	1.49	1.50	1.50
–0.40	1.49	1.49	1.49	1.49	1.49	1.49	1.49	1.49	1.49	1.49
–0.30	1.47	1.48	1.48	1.48	1.48	1.48	1.48	1.48	1.48	1.48
–0.20	1.46	1.46	1.46	1.46	1.47	1.47	1.47	1.47	1.47	1.47
–0.10	1.43	1.44	1.44	1.44	1.45	1.45	1.45	1.46	1.46	1.46
0.00	1.40	1.41	1.41	1.42	1.42	1.43	1.43	1.43	1.44	1.44
0.10	1.36	1.37	1.38	1.39	1.39	1.40	1.40	1.41	1.41	1.42
0.20	1.32	1.33	1.34	1.35	1.36	1.37	1.37	1.38	1.38	1.39
0.30	1.28	1.29	1.30	1.31	1.32	1.33	1.34	1.34	1.35	1.36
0.40	1.23	1.24	1.26	1.27	1.28	1.29	1.30	1.31	1.32	1.32
0.50	1.18	1.20	1.21	1.23	1.24	1.25	1.26	1.27	1.28	1.29
0.60	1.14	1.16	1.17	1.19	1.20	1.21	1.23	1.24	1.25	1.26
0.70	1.11	1.12	1.14	1.15	1.17	1.18	1.19	1.20	1.21	1.22
0.80	1.08	1.09	1.11	1.12	1.13	1.15	1.16	1.17	1.18	1.19
0.90	1.05	1.07	1.08	1.09	1.10	1.12	1.13	1.14	1.15	1.16
1.00	1.04	1.05	1.06	1.07	1.08	1.09	1.10	1.12	1.13	1.14
1.10	1.02	1.03	1.04	1.05	1.06	1.07	1.08	1.09	1.10	1.12
1.20	1.01	1.02	1.03	1.03	1.04	1.05	1.06	1.07	1.08	1.09
1.30	1.01	1.01	1.02	1.02	1.03	1.04	1.05	1.06	1.07	1.08
1.40	1.00	1.01	1.01	1.01	1.02	1.03	1.03	1.04	1.05	1.06
1.50	1.00	1.00	1.01	1.01	1.01	1.02	1.02	1.03	1.04	1.04
1.60	1.00	1.00	1.00	1.00	1.01	1.01	1.02	1.02	1.03	1.03
1.70	1.00	1.00	1.00	1.00	1.00	1.01	1.01	1.01	1.02	1.02
1.80	1.00	1.00	1.00	1.00	1.00	1.00	1.01	1.01	1.01	1.02
1.90	1.00	1.00	1.00	1.00	1.00	1.00	1.00	1.01	1.01	1.01
2.00	1.00	1.00	1.00	1.00	1.00	1.00	1.00	1.00	1.00	1.01

Table 27: The Effective Cost of Production for a one-sided specification with stated target when nonconforming product is reworked and the cost of rework is 0.33 of the nominal cost of production

	Tolerance in Standard Deviation Units										
C_{pk}	***0.0***	***0.3***	***0.6***	***0.9***	***1.2***	***1.5***	***1.8***	***2.1***	***2.4***	***2.7***	***3.0***
–0.50	1.31	1.31	1.32	1.32	1.32	1.32	1.32	1.33	1.33	1.33	1.33
–0.40	1.30	1.30	1.31	1.31	1.31	1.32	1.32	1.32	1.32	1.32	1.32
–0.30	1.27	1.28	1.29	1.29	1.30	1.30	1.31	1.31	1.31	1.31	1.31
–0.20	1.24	1.25	1.26	1.27	1.28	1.28	1.29	1.29	1.30	1.30	1.30
–0.10	1.21	1.22	1.23	1.24	1.25	1.26	1.27	1.27	1.28	1.28	1.29
0.00	1.17	1.18	1.19	1.21	1.22	1.23	1.24	1.24	1.25	1.26	1.26
0.10	1.13	1.14	1.15	1.17	1.18	1.19	1.20	1.21	1.22	1.23	1.24
0.20	1.09	1.10	1.12	1.13	1.14	1.15	1.17	1.18	1.19	1.20	1.21
0.30	1.06	1.07	1.08	1.09	1.10	1.12	1.13	1.14	1.15	1.16	1.17
0.40	1.04	1.05	1.05	1.06	1.07	1.08	1.10	1.11	1.12	1.13	1.14
0.50	1.02	1.03	1.03	1.04	1.05	1.06	1.07	1.08	1.09	1.10	1.11
0.60	1.01	1.02	1.02	1.02	1.03	1.04	1.05	1.05	1.06	1.07	1.09
0.70	1.01	1.01	1.01	1.01	1.02	1.02	1.03	1.04	1.04	1.05	1.06
0.80	1.00	1.00	1.00	1.01	1.01	1.01	1.02	1.02	1.03	1.04	1.04
0.90	1.00	1.00	1.00	1.00	1.00	1.01	1.01	1.01	1.02	1.02	1.03
1.00	1.00	1.00	1.00	1.00	1.00	1.00	1.00	1.01	1.01	1.01	1.02
1.10	1.00	1.00	1.00	1.00	1.00	1.00	1.00	1.00	1.01	1.01	1.01
1.20	1.00	1.00	1.00	1.00	1.00	1.00	1.00	1.00	1.00	1.00	1.01
1.30	1.00	1.00	1.00	1.00	1.00	1.00	1.00	1.00	1.00	1.00	1.00
1.40	1.00	1.00	1.00	1.00	1.00	1.00	1.00	1.00	1.00	1.00	1.00

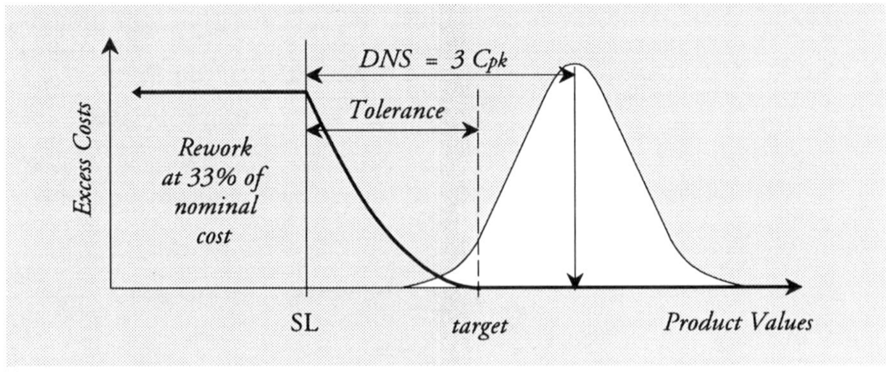

Table 27: One-Sided Specification With a Stated Target Value

Table 27: The Effective Cost of Production for a one-sided specification with stated target when nonconforming product is reworked and the cost of rework is 0.33 of the nominal cost of production

Tolerance in Standard Deviation Units

C_{pk}	3.3	3.6	3.9	4.2	4.5	4.8	5.1	5.4	5.7	6.0
–0.50	1.33	1.33	1.33	1.33	1.33	1.33	1.33	1.33	1.33	1.33
–0.40	1.32	1.32	1.32	1.33	1.33	1.33	1.33	1.33	1.33	1.33
–0.30	1.32	1.32	1.32	1.32	1.32	1.32	1.32	1.32	1.32	1.32
–0.20	1.30	1.31	1.31	1.31	1.31	1.31	1.31	1.31	1.32	1.32
–0.10	1.29	1.29	1.29	1.30	1.30	1.30	1.30	1.30	1.30	1.31
0.00	1.27	1.27	1.28	1.28	1.28	1.28	1.29	1.29	1.29	1.29
0.10	1.24	1.25	1.25	1.26	1.26	1.27	1.27	1.27	1.27	1.28
0.20	1.21	1.22	1.23	1.23	1.24	1.24	1.25	1.25	1.26	1.26
0.30	1.18	1.19	1.20	1.21	1.21	1.22	1.22	1.23	1.23	1.24
0.40	1.15	1.16	1.17	1.18	1.19	1.19	1.20	1.21	1.21	1.22
0.50	1.12	1.13	1.14	1.15	1.16	1.17	1.17	1.18	1.19	1.19
0.60	1.10	1.11	1.12	1.13	1.13	1.14	1.15	1.16	1.16	1.17
0.70	1.07	1.08	1.09	1.10	1.11	1.12	1.13	1.14	1.14	1.15
0.80	1.05	1.06	1.07	1.08	1.09	1.10	1.11	1.11	1.12	1.13
0.90	1.04	1.04	1.05	1.06	1.07	1.08	1.09	1.09	1.10	1.11
1.00	1.02	1.03	1.04	1.05	1.05	1.06	1.07	1.08	1.08	1.09
1.10	1.02	1.02	1.03	1.03	1.04	1.05	1.05	1.06	1.17	1.08
1.20	1.01	1.01	1.02	1.02	1.03	1.03	1.04	1.05	1.16	1.06
1.30	1.01	1.01	1.01	1.01	1.02	1.02	1.03	1.04	1.14	1.05
1.40	1.00	1.00	1.01	1.01	1.01	1.02	1.02	1.03	1.03	1.04
1.50	1.00	1.00	1.00	1.01	1.01	1.01	1.02	1.02	1.02	1.03
1.60	1.00	1.00	1.00	1.00	1.00	1.01	1.01	1.01	1.02	1.02
1.70	1.00	1.00	1.00	1.00	1.00	1.00	1.01	1.01	1.01	1.02
1.80	1.00	1.00	1.00	1.00	1.00	1.00	1.00	1.01	1.01	1.01
1.90	1.00	1.00	1.00	1.00	1.00	1.00	1.00	1.00	1.01	1.01
2.00	1.00	1.00	1.00	1.00	1.00	1.00	1.00	1.00	1.00	1.00

Table 28: The Effective Cost of Production for a one-sided specification with stated target when nonconforming product is reworked and the cost of rework is 0.20 of the nominal cost of production

	Tolerance in Standard Deviation Units										
C_{pk}	***0.0***	***0.3***	***0.6***	***0.9***	***1.2***	***1.5***	***1.8***	***2.1***	***2.4***	***2.7***	***3.0***
–0.50	1.19	1.19	1.19	1.19	1.19	1.19	1.19	1.20	1.20	1.20	1.20
–0.40	1.18	1.18	1.18	1.19	1.19	1.19	1.19	1.19	1.19	1.19	1.19
–0.30	1.16	1.17	1.17	1.18	1.18	1.18	1.18	1.19	1.19	1.19	1.19
–0.20	1.15	1.15	1.16	1.16	1.17	1.17	1.17	1.18	1.18	1.18	1.18
–0.10	1.12	1.13	1.14	1.14	1.15	1.15	1.16	1.16	1.17	1.17	1.17
0.00	1.10	1.11	1.12	1.12	1.13	1.14	1.14	1.15	1.15	1.15	1.16
0.10	1.08	1.08	1.09	1.10	1.11	1.11	1.12	1.13	1.13	1.14	1.14
0.20	1.05	1.06	1.07	1.08	1.08	1.09	1.10	1.11	1.11	1.12	1.12
0.30	1.04	1.04	1.05	1.06	1.06	1.07	1.08	1.09	1.09	1.10	1.10
0.40	1.02	1.03	1.03	1.04	1.04	1.05	1.06	1.06	1.07	1.08	1.09
0.50	1.01	1.02	1.02	1.02	1.03	1.03	1.04	1.05	1.05	1.06	1.07
0.60	1.01	1.01	1.01	1.01	1.02	1.02	1.03	1.03	1.04	1.04	1.05
0.70	1.00	1.00	1.01	1.01	1.01	1.01	1.02	1.02	1.03	1.03	1.04
0.80	1.00	1.00	1.00	1.00	1.01	1.01	1.01	1.01	1.02	1.02	1.03
0.90	1.00	1.00	1.00	1.00	1.00	1.00	1.01	1.01	1.01	1.01	1.02
1.00	1.00	1.00	1.00	1.00	1.00	1.0	1.00	1.00	1.01	1.01	1.01
1.10	1.00	1.00	1.00	1.00	1.00	1.00	1.00	1.00	1.00	1.00	1.01
1.20	1.00	1.00	1.00	1.00	1.00	1.00	1.00	1.00	1.00	1.00	1.00
1.30	1.00	1.00	1.00	1.00	1.00	1.00	1.00	1.00	1.00	1.00	1.00

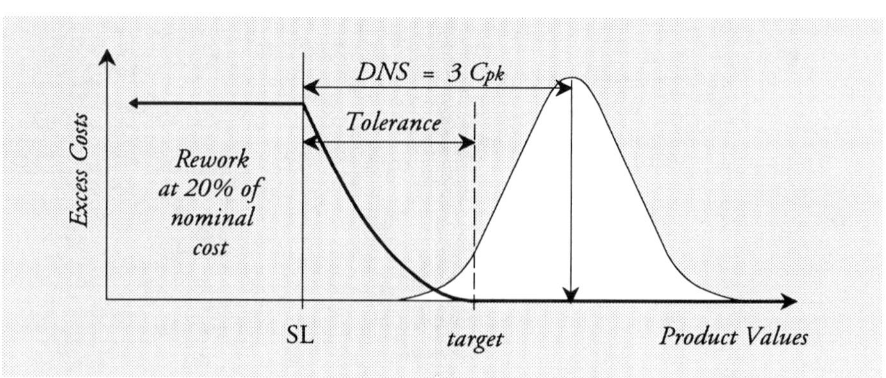

Table 28: One-Sided Specification With a Stated Target Value

Table 28: The Effective Cost of Production for a one-sided specification with stated target when nonconforming product is reworked and the cost of rework is 0.20 of the nominal cost of production

Tolerance in Standard Deviation Units

C_{pk}	3.3	3.6	3.9	4.2	4.5	4.8	5.1	5.4	5.7	6.0
–0.50	1.20	1.20	1.20	1.20	1.20	1.20	1.20	1.20	1.20	1.20
–0.40	1.19	1.19	1.19	1.20	1.20	1.20	1.20	1.20	1.20	1.20
–0.30	1.19	1.19	1.19	1.19	1.19	1.19	1.19	1.19	1.19	1.19
–0.20	1.18	1.18	1.18	1.19	1.19	1.19	1.19	1.19	1.19	1.19
–0.10	1.17	1.17	1.18	1.18	1.18	1.18	1.18	1.18	1.18	1.18
0.00	1.16	1.16	1.17	1.17	1.17	1.17	1.17	1.17	1.17	1.18
0.10	1.15	1.15	1.15	1.15	1.16	1.16	1.16	1.16	1.16	1.17
0.20	1.13	1.13	1.14	1.14	1.14	1.15	1.15	1.15	1.15	1.16
0.30	1.11	1.12	1.12	1.12	1.13	1.13	1.13	1.14	1.14	1.14
0.40	1.09	1.10	1.10	1.11	1.11	1.12	1.12	1.12	1.13	1.13
0.50	1.07	1.08	1.09	1.09	1.10	1.10	1.10	1.11	1.11	1.12
0.60	1.06	1.06	1.07	1.08	1.08	1.09	1.09	1.09	1.10	1.10
0.70	1.04	1.05	1.05	1.06	1.07	1.07	1.08	1.08	1.09	1.09
0.80	1.03	1.04	1.04	1.05	1.05	1.06	1.06	1.07	1.07	1.08
0.90	1.02	1.03	1.03	1.04	1.04	1.05	1.05	1.06	1.06	1.07
1.00	1.01	1.02	1.02	1.03	1.03	1.04	1.04	1.05	1.05	1.06
1.10	1.01	1.01	1.02	1.02	1.02	1.03	1.03	1.04	1.04	1.05
1.20	1.01	1.01	1.01	1.01	1.02	1.02	1.02	1.03	1.03	1.04
1.30	1.00	1.00	1.01	1.01	1.01	1.01	1.02	1.02	1.03	1.03
1.40	1.00	1.00	1.00	1.01	1.01	1.01	1.01	1.02	1.02	1.02
1.50	1.00	1.00	1.00	1.00	1.00	1.01	1.01	1.01	1.01	1.02
1.60	1.00	1.00	1.00	1.00	1.00	1.00	1.01	1.01	1.01	1.01
1.70	1.00	1.00	1.00	1.00	1.00	1.00	1.00	1.01	1.01	1.01
1.80	1.00	1.00	1.00	1.00	1.00	1.00	1.00	1.00	1.00	1.01
1.90	1.00	1.00	1.00	1.00	1.00	1.00	1.00	1.00	1.00	1.00

Table 29: The Effective Cost of Production for a one-sided specification with stated target when nonconforming product is reworked and the cost of rework is 0.10 of the nominal cost of production

	Tolerance in Standard Deviation Units										
C_{pk}	*0.0*	*0.3*	*0.6*	*0.9*	*1.2*	*1.5*	*1.8*	*2.1*	*2.4*	*2.7*	*3.0*
–0.50	1.09	1.09	1.10	1.10	1.10	1.10	1.10	1.10	1.10	1.10	1.10
–0.40	1.09	1.09	1.09	1.09	1.09	1.09	1.10	1.10	1.10	1.10	1.10
–0.30	1.08	1.08	1.09	1.09	1.09	1.09	1.09	1.09	1.09	1.09	1.09
–0.20	1.07	1.08	1.08	1.08	1.08	1.09	1.09	1.09	1.09	1.09	1.09
–0.10	1.06	1.07	1.07	1.07	1.07	1.08	1.08	1.08	1.08	1.08	1.09
0.00	1.05	1.05	1.06	1.06	1.06	1.07	1.07	1.07	1.08	1.08	1.08
0.10	1.04	1.04	1.05	1.05	1.05	1.06	1.06	1.06	1.07	1.07	1.07
0.20	1.03	1.03	1.03	1.04	1.04	1.05	1.05	1.05	1.06	1.06	1.06
0.30	1.02	1.02	1.02	1.03	1.03	1.04	1.04	1.04	1.05	1.05	1.05
0.40	1.01	1.01	1.02	1.02	1.02	1.03	1.03	1.03	1.04	1.04	1.04
0.50	1.01	1.01	1.01	1.01	1.01	1.02	1.02	1.02	1.03	1.03	1.03
0.60	1.00	1.00	1.01	1.01	1.01	1.01	1.01	1.02	1.02	1.02	1.03
0.70	1.00	1.00	1.00	1.00	1.01	1.01	1.01	1.01	1.01	1.02	1.02
0.80	1.00	1.00	1.00	1.00	1.00	1.00	1.00	1.01	1.01	1.01	1.01
0.90	1.00	1.00	1.00	1.00	1.00	1.00	1.00	1.00	1.01	1.01	1.01
1.00	1.00	1.00	1.00	1.00	1.00	1.00	1.00	1.00	1.00	1.00	1.01
1.10	1.00	1.00	1.00	1.00	1.00	1.00	1.00	1.00	1.00	1.00	1.00

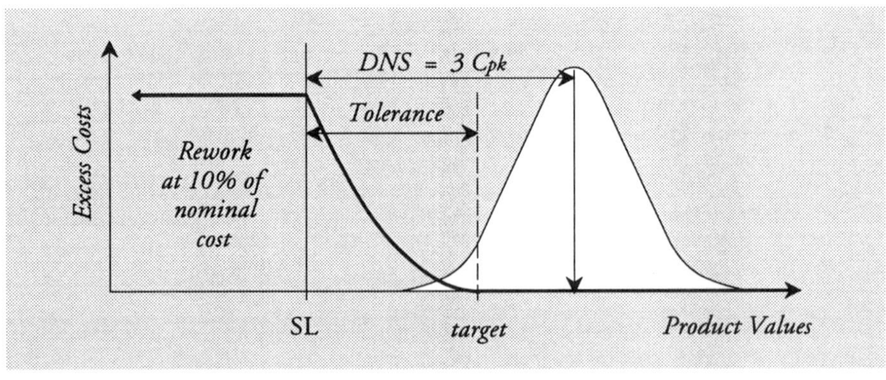

Table 29: One-Sided Specification With a Stated Target Value

Table 29: The Effective Cost of Production for a one-sided specification with stated target when nonconforming product is reworked and the cost of rework is 0.10 of the nominal cost of production

	Tolerance in Standard Deviation Units									
C_{pk}	***3.3***	***3.6***	***3.9***	***4.2***	***4.5***	***4.8***	***5.1***	***5.4***	***5.7***	***6.0***
–0.50	1.10	1.10	1.10	1.10	1.10	1.10	1.10	1.10	1.10	1.10
–0.40	1.10	1.10	1.10	1.10	1.10	1.10	1.10	1.10	1.10	1.10
–0.30	1.09	1.10	1.10	1.10	1.10	1.10	1.10	1.10	1.10	1.10
–0.20	1.09	1.09	1.09	1.09	1.09	1.09	1.09	1.09	1.09	1.09
–0.10	1.09	1.09	1.09	1.09	1.09	1.09	1.09	1.09	1.09	1.09
0.00	1.08	1.08	1.08	1.08	1.08	1.09	1.09	1.09	1.09	1.09
0.10	1.07	1.07	1.08	1.08	1.08	1.08	1.08	1.08	1.08	1.08
0.20	1.06	1.07	1.07	1.07	1.07	1.07	1.07	1.08	1.08	1.08
0.30	1.06	1.06	1.06	1.06	1.06	1.07	1.07	1.07	1.07	1.07
0.40	1.05	1.05	1.05	1.05	1.06	1.06	1.06	1.06	1.06	1.06
0.50	1.04	1.04	1.04	1.05	1.05	1.05	1.05	1.05	1.06	1.06
0.60	1.03	1.03	1.03	1.04	1.04	1.04	1.05	1.05	1.05	1.05
0.70	1.02	1.02	1.03	1.03	1.03	1.04	1.04	1.04	1.04	1.04
0.80	1.01	1.02	1.02	1.02	1.03	1.03	1.03	1.03	1.04	1.04
0.90	1.01	1.01	1.02	1.02	1.02	1.02	1.03	1.03	1.03	1.03
1.00	1.00	1.01	1.01	1.01	1.02	1.02	1.02	1.02	1.03	1.03
1.10	1.00	1.01	1.01	1.01	1.01	1.01	1.02	1.02	1.02	1.02
1.20	1.00	1.00	1.01	1.01	1.01	1.01	1.01	1.01	1.02	1.02
1.30	1.00	1.00	1.00	1.00	1.01	1.01	1.01	1.01	1.01	1.02
1.40	1.00	1.00	1.00	1.00	1.00	1.01	1.01	1.01	1.01	1.01
1.50	1.00	1.00	1.00	1.00	1.00	1.00	1.00	1.01	1.01	1.01
1.60	1.00	1.00	1.00	1.00	1.00	1.00	1.00	1.00	1.01	1.01
1.70	1.00	1.00	1.00	1.00	1.00	1.00	1.00	1.00	1.00	1.00

Tables 30 to 37

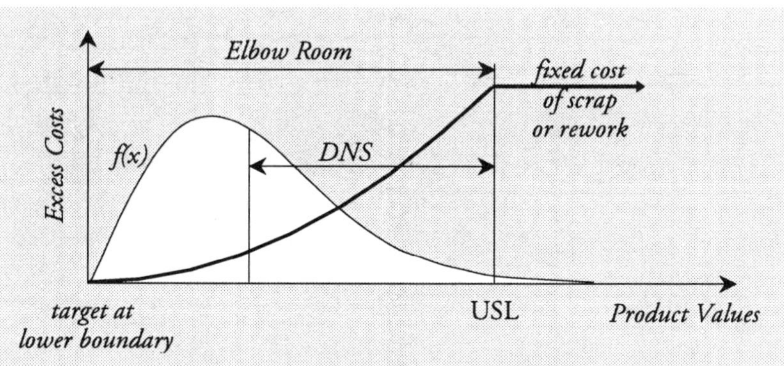

Tables 30 to 37: The Excess Cost Function When Target is at a Lower Boundary

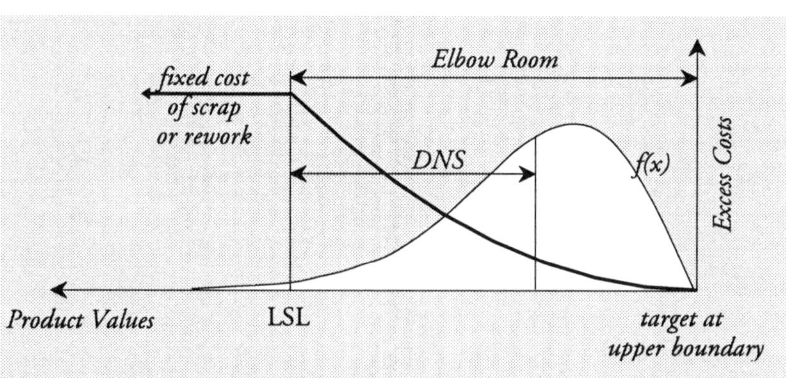

Tables 30 to 37: The Excess Cost Function When Target is at an Upper Boundary

Tables 30 to 37
The Effective Cost of Production When the Target is at a Boundary

In Tables 30 to 37, as the *DNS* value approaches the *Elbow Room* value the process average approaches the target value.

Blanks in Tables 30 to 37 correspond to impossible conditions.

Table 30: The Effective Cost of Production when the target is at a boundary and nonconforming units are scrapped

Elbow Room in standard deviation units

DNS	0.3	0.6	0.9	1.2	1.5	1.8	2.1	2.4	2.7	3.0
0.0	1.328	1.561	1.775	1.954	2.11	2.23	2.33	2.42	2.50	2.56
0.3	1.000	1.198	1.387	1.556	1.700	1.818	1.917	1.999	2.07	2.12
0.6	-	1.000	1.141	1.290	1.424	1.538	1.632	1.710	1.775	1.830
0.9	-	-	1.000	1.109	1.228	1.335	1.427	1.503	1.566	1.620
1.2	-	-	-	1.000	1.088	1.185	1.272	1.347	1.409	1.462
1.5	-	-	-	-	1.000	1.073	1.154	1.225	1.287	1.339
1.8	-	-	-	-	-	1.000	1.062	1.130	1.189	1.241
2.1	-	-	-	-	-	-	1.000	1.054	1.112	1.161
2.4	-	-	-	-	-	-	-	1.000	1.047	1.097
2.7	-	-	-	-	-	-	-	-	1.000	1.042
3.0	-	-	-	-	-	-	-	-	-	1.000

Elbow Room in standard deviation units

DNS	3.3	3.6	3.9	4.2	4.5	4.8	5.1	5.4	5.7	6.0
0.0	2.61	2.66	2.70	2.74	2.77	2.80	2.83	2.85	2.87	2.90
0.3	2.18	2.22	2.25	2.29	2.32	2.34	2.37	2.39	2.41	2.42
0.6	1.876	1.917	1.951	1.981	2.01	2.03	2.05	2.07	2.09	2.11
0.9	1.665	1.703	1.738	1.767	1.793	1.816	1.837	1.855	1.872	1.887
1.2	1.507	1.546	1.579	1.609	1.636	1.659	1.681	1.700	1.717	1.733
1.5	1.384	1.423	1.457	1.488	1.515	1.540	1.563	1.583	1.602	1.619
1.8	1.285	1.324	1.359	1.390	1.418	1.444	1.468	1.490	1.511	1.530
2.1	1.205	1.243	1.277	1.308	1.337	1.364	1.389	1.412	1.434	1.455
2.4	1.139	1.176	1.209	1.239	1.268	1.295	1.321	1.345	1.368	1.389
2.7	1.085	1.120	1.153	1.181	1.209	1.235	1.261	1.285	1.308	1.331
3.0	1.037	1.075	1.105	1.133	1.159	1.184	1.208	1.232	1.255	1.278
3.3	1.000	1.033	1.066	1.093	1.117	1.140	1.163	1.186	1.208	1.230
3.6	-	1.000	1.030	1.059	1.082	1.104	1.124	1.145	1.167	1.188
3.9	-	-	1.000	1.027	1.053	1.073	1.092	1.111	1.130	1.150
4.2	-	-	-	1.000	1.025	1.048	1.066	1.083	1.100	1.118
4.5	-	-	-	-	1.000	1.023	1.043	1.059	1.074	1.090
4.8	-	-	-	-	-	1.000	1.021	1.040	1.054	1.067
5.1	-	-	-	-	-	-	1.000	1.020	1.036	1.049
5.4	-	-	-	-	-	-	-	1.000	1.018	1.033
5.7	-	-	-	-	-	-	-	-	1.000	1.017
6.0	-	-	-	-	-	-	-	-	-	1.000

Table 30: Target at boundary, nonconforming units scrapped

	Elbow Room in standard deviation units									
DNS	*6.3*	*6.6*	*6.9*	*7.2*	*7.5*	*7.8*	*8.1*	*8.4*	*8.7*	*9.0*
0.0	2.91	2.93	2.94	2.96	2.97	2.98	2.99	3.01	3.02	3.03
0.3	2.44	2.45	2.47	2.48	2.49	2.50	2.51	2.52	2.53	2.54
0.6	2.12	2.13	2.14	2.16	2.17	2.18	2.18	2.19	2.20	2.21
0.9	1.901	1.914	1.926	1.936	1.946	1.956	1.964	1.972	1.980	1.987
1.2	1.748	1.761	1.773	1.785	1.796	1.805	1.815	1.823	1.831	1.839
1.5	1.635	1.650	1.663	1.676	1.688	1.699	1.709	1.719	1.728	1.737
1.8	1.547	1.563	1.579	1.593	1.607	1.619	1.631	1.642	1.653	1.662
2.1	1.474	1.492	1.509	1.525	1.540	1.554	1.567	1.580	1.592	1.603
2.4	1.410	1.429	1.448	1.465	1.481	1.497	1.511	1.525	1.538	1.551
2.7	1.352	1.373	1.392	1.410	1.428	1.444	1.460	1.475	1.489	1.502
3.0	1.300	1.321	1.341	1.360	1.378	1.395	1.412	1.427	1.442	1.457
3.3	1.252	1.273	1.293	1.313	1.331	1.349	1.366	1.383	1.398	1.413
3.6	1.209	1.230	1.250	1.269	1.288	1.306	1.324	1.341	1.357	1.372
3.9	1.170	1.190	1.210	1.229	1.248	1.266	1.284	1.301	1.318	1.333
4.2	1.136	1.155	1.174	1.193	1.211	1.229	1.247	1.264	1.281	1.297
4.5	1.107	1.124	1.142	1.160	1.178	1.195	1.213	1.230	1.246	1.262
4.8	1.082	1.097	1.114	1.130	1.147	1.164	1.181	1.198	1.214	1.230
5.1	1.061	1.075	1.089	1.104	1.120	1.136	1.152	1.169	1.184	1.200
5.4	1.044	1.056	1.068	1.082	1.096	1.111	1.126	1.142	1.157	1.172
5.7	1.030	1.041	1.051	1.063	1.075	1.089	1.103	1.117	1.132	1.147
6.0	1.016	1.028	1.037	1.047	1.058	1.070	1.082	1.096	1.110	1.123
6.3	1.000	1.015	1.026	1.034	1.043	1.053	1.065	1.077	1.089	1.102
6.6	-	1.000	1.014	1.024	1.032	1.040	1.050	1.060	1.071	1.083
6.9	-	-	1.000	1.013	1.022	1.029	1.037	1.046	1.056	1.067
7.2	-	-	-	1.000	1.012	1.021	1.027	1.035	1.043	1.052
7.5	-	-	-	-	1.000	1.012	1.019	1.025	1.032	1.040
7.8	-	-	-	-	-	1.000	1.011	1.018	1.024	1.030
8.1	-	-	-	-	-	-	1.000	1.010	1.017	1.022
8.4	-	-	-	-	-	-	-	1.000	1.010	1.016
8.7	-	-	-	-	-	-	-	-	1.000	1.009
9.0	-	-	-	-	-	-	-	-	-	1.000

If you have a negative *DNS* value, use Table 1 to obtain an approximate Effective Cost of Production. To do this divide the *DNS* value by 3 and divide the *Elbow Room* value by 3 (yes 3) to obtain the corresponding values for C_{pk} and C_p to use with Table 1 and look up the Effective Cost.

Table 30: Target at boundary, nonconforming units scrapped

	Elbow Room in standard deviation units									
DNS	***9.3***	***9.6***	***9.9***	***10.2***	***10.5***	***10.8***	***11.1***	***11.4***	***11.7***	***12.0***
0.0	3.03	3.04	3.05	3.06	3.07	3.07	3.08	3.08	3.09	3.10
0.3	2.54	2.55	2.56	2.56	2.57	2.57	2.58	2.58	2.59	2.59
0.6	2.21	2.22	2.23	2.23	2.24	2.24	2.25	2.25	2.26	2.26
0.9	1.994	2.00	2.01	2.01	2.02	2.02	2.03	2.03	2.04	2.04
1.2	1.846	1.853	1.859	1.865	1.871	1.877	1.882	1.887	1.891	1.896
1.5	1.745	1.753	1.760	1.767	1.774	1.780	1.786	1.791	1.797	1.802
1.8	1.672	1.681	1.689	1.697	1.705	1.712	1.719	1.726	1.732	1.738
2.1	1.614	1.624	1.633	1.643	1.651	1.660	1.668	1.675	1.682	1.689
2.4	1.563	1.574	1.585	1.595	1.605	1.614	1.623	1.631	1.639	1.647
2.7	1.515	1.528	1.539	1.550	1.561	1.571	1.581	1.590	1.599	1.608
3.0	1.470	1.484	1.496	1.508	1.519	1.530	1.541	1.551	1.560	1.569
3.3	1.428	1.442	1.455	1.467	1.479	1.491	1.502	1.513	1.523	1.533
3.6	1.387	1.401	1.415	1.428	1.441	1.453	1.465	1.476	1.487	1.497
3.9	1.349	1.363	1.378	1.391	1.404	1.417	1.429	1.441	1.452	1.463
4.2	1.312	1.327	1.342	1.356	1.369	1.382	1.395	1.407	1.418	1.429
4.5	1.278	1.293	1.308	1.322	1.336	1.349	1.362	1.374	1.386	1.398
4.8	1.246	1.261	1.276	1.290	1.304	1.317	1.330	1.343	1.355	1.367
5.1	1.216	1.231	1.245	1.260	1.274	1.287	1.300	1.313	1.326	1.338
5.4	1.187	1.202	1.217	1.231	1.245	1.259	1.272	1.285	1.297	1.309
5.7	1.161	1.176	1.190	1.204	1.218	1.232	1.245	1.258	1.270	1.283
6.0	1.137	1.151	1.165	1.179	1.193	1.206	1.219	1.232	1.245	1.257
6.3	1.116	1.129	1.142	1.156	1.169	1.182	1.195	1.208	1.220	1.233
6.6	1.096	1.109	1.121	1.134	1.147	1.160	1.172	1.185	1.197	1.209
6.9	1.078	1.090	1.102	1.114	1.127	1.139	1.151	1.164	1.176	1.188
7.2	1.063	1.073	1.085	1.096	1.108	1.120	1.132	1.143	1.155	1.167
7.5	1.049	1.059	1.069	1.080	1.091	1.102	1.113	1.125	1.136	1.148
7.8	1.038	1.046	1.055	1.065	1.075	1.086	1.097	1.107	1.118	1.129
8.1	1.028	1.035	1.043	1.052	1.061	1.071	1.081	1.091	1.102	1.113
8.4	1.021	1.026	1.033	1.041	1.049	1.058	1.067	1.077	1.087	1.097
8.7	1.015	1.020	1.025	1.031	1.038	1.046	1.055	1.064	1.073	1.083
9.0	1.009	1.014	1.018	1.023	1.029	1.036	1.044	1.052	1.061	1.069
9.3	1.000	1.008	1.013	1.017	1.022	1.028	1.034	1.042	1.049	1.058
9.6	-	1.000	1.008	1.013	1.016	1.021	1.026	1.033	1.040	1.047
9.9	-	-	1.000	1.008	1.012	1.015	1.020	1.025	1.031	1.038
10.2	-	-	-	1.000	1.007	1.011	1.015	1.019	1.024	1.029
10.5	-	-	-	-	1.000	1.007	1.011	1.014	1.018	1.023
10.8	-	-	-	-	-	1.000	1.007	1.010	1.013	1.017
11.1	-	-	-	-	-	-	1.000	1.006	1.010	1.013
11.4	-	-	-	-	-	-	-	1.000	1.006	1.009
11.7	-	-	-	-	-	-	-	-	1.000	1.006

Table 30: Target at boundary, nonconforming product scrapped

DNS	Elbow Room in std. dev. units				
	15.	18.	24.	36.	60.
0.0	2.90	2.92	2.94	2.96	2.98
0.3	2.50	2.52	2.55	2.57	2.59
0.6	2.24	2.27	2.29	2.32	2.34
0.9	2.07	2.10	2.13	2.16	2.19
1.2	1.95	1.98	2.02	2.05	2.08
1.5	1.87	1.90	1.94	1.98	2.02
1.8	1.81	1.84	1.89	1.94	1.98
2.1	1.76	1.80	1.85	1.90	1.95
2.4	1.72	1.76	1.82	1.88	1.93
2.7	1.68	1.73	1.79	1.86	1.92
3.0	1.65	1.70	1.77	1.84	1.90
3.3	1.61	1.67	1.75	1.83	1.89
3.6	1.58	1.64	1.72	1.81	1.88
3.9	1.55	1.62	1.70	1.80	1.87
4.2	1.52	1.59	1.68	1.78	1.87
4.5	1.49	1.57	1.66	1.77	1.86
4.8	1.47	1.54	1.64	1.75	1.85
5.1	1.44	1.52	1.62	1.74	1.84
5.4	1.41	1.49	1.60	1.72	1.83
5.7	1.39	1.47	1.58	1.71	1.82
6.0	1.36	1.45	1.56	1.70	1.81
7.5	1.25	1.34	1.47	1.63	1.77
9.0	1.16	1.25	1.39	1.56	1.72
12.	1.04	1.11	1.25	1.45	1.64
15.	1.00	1.03	1.14	1.34	1.56
18.	-	1.00	1.06	1.25	1.49
24.	-	-	1.00	1.11	1.36
36.	-	-	-	1.00	1.16
60.	-	-	-	-	1.00

Table 31: The Effective Cost of Production when the target is at a boundary and nonconforming units are reworked

Elbow Room in standard deviation units

DNS	*0.3*	*0.6*	*0.9*	*1.2*	*1.5*	*1.8*	*2.1*	*2.4*	*2.7*	*3.0*
0.0	1.264	1.400	1.503	1.577	1.635	1.679	1.714	1.742	1.766	1.785
0.3	1.000	1.175	1.309	1.414	1.493	1.554	1.603	1.643	1.675	1.702
0.6	-	1.000	1.130	1.248	1.344	1.420	1.481	1.531	1.572	1.607
0.9	-	-	1.000	1.103	1.204	1.288	1.358	1.416	1.464	1.505
1.2	-	-	-	1.000	1.084	1.170	1.244	1.307	1.359	1.405
1.5	-	-	-	-	1.000	1.071	1.144	1.208	1.263	1.310
1.8	-	-	-	-	-	1.000	1.061	1.124	1.178	1.227
2.1	-	-	-	-	-	-	1.000	1.053	1.107	1.154
2.4	-	-	-	-	-	-	-	1.000	1.046	1.094
2.7	-	-	-	-	-	-	-	-	1.000	1.041
3.0	-	-	-	-	-	-	-	-	-	1.000

Elbow Room in standard deviation units

DNS	*3.3*	*3.6*	*3.9*	*4.2*	*4.5*	*4.8*	*5.1*	*5.4*	*5.7*	*6.0*
0.0	1.802	1.816	1.828	1.839	1.849	1.857	1.865	1.872	1.878	1.884
0.3	1.725	1.745	1.762	1.777	1.790	1.802	1.812	1.822	1.831	1.838
0.6	1.636	1.662	1.684	1.703	1.721	1.736	1.750	1.762	1.773	1.784
0.9	1.541	1.571	1.598	1.622	1.643	1.662	1.679	1.694	1.708	1.721
1.2	1.444	1.479	1.509	1.537	1.561	1.583	1.603	1.621	1.638	1.653
1.5	1.352	1.389	1.422	1.452	1.479	1.504	1.527	1.547	1.566	1.584
1.8	1.269	1.307	1.341	1.372	1.401	1.428	1.452	1.475	1.496	1.515
2.1	1.196	1.233	1.268	1.299	1.329	1.356	1.382	1.406	1.428	1.449
2.4	1.134	1.171	1.204	1.234	1.264	1.291	1.317	1.342	1.365	1.387
2.7	1.082	1.117	1.149	1.178	1.206	1.233	1.259	1.284	1.307	1.330
3.0	1.037	1.073	1.103	1.131	1.157	1.182	1.207	1.231	1.255	1.278
3.3	1.000	1.033	1.065	1.091	1.116	1.139	1.162	1.185	1.208	1.230
3.6	-	1.000	1.030	1.058	1.081	1.103	1.124	1.145	1.166	1.188
3.9	-	-	1.000	1.027	1.052	1.073	1.092	1.111	1.130	1.150
4.2	-	-	-	1.000	1.025	1.047	1.065	1.082	1.100	1.118
4.5	-	-	-	-	1.000	1.023	1.043	1.059	1.074	1.090
4.8	-	-	-	-	-	1.000	1.021	1.039	1.053	1.067
5.1	-	-	-	-	-	-	1.000	1.020	1.036	1.049
5.4	-	-	-	-	-	-	-	1.000	1.018	1.033
5.7	-	-	-	-	-	-	-	-	1.000	1.017
6.0	-	-	-	-	-	-	-	-	-	1.000

Table 31: Target at boundary, rework at 100% of nominal

DNS	Elbow Room in standard deviation units									
	6.3	6.6	6.9	7.2	7.5	7.8	8.1	8.4	8.7	9.0
0.0	1.889	1.893	1.898	1.902	1.905	1.909	1.912	1.915	1.918	1.921
0.3	1.845	1.852	1.858	1.864	1.869	1.873	1.878	1.882	1.886	1.890
0.6	1.793	1.802	1.809	1.817	1.824	1.830	1.836	1.841	1.846	1.851
0.9	1.733	1.743	1.754	1.763	1.771	1.779	1.787	1.794	1.800	1.806
1.2	1.667	1.680	1.692	1.704	1.714	1.724	1.733	1.741	1.749	1.757
1.5	1.600	1.615	1.629	1.642	1.654	1.665	1.676	1.686	1.696	1.704
1.8	1.533	1.550	1.566	1.580	1.594	1.607	1.619	1.631	1.641	1.652
2.1	1.469	1.487	1.504	1.521	1.536	1.550	1.564	1.576	1.589	1.600
2.4	1.408	1.428	1.446	1.464	1.480	1.496	1.510	1.524	1.538	1.550
2.7	1.352	1.373	1.392	1.410	1.427	1.444	1.460	1.475	1.489	1.502
3.0	1.300	1.321	1.341	1.360	1.378	1.395	1.412	1.427	1.442	1.457
3.3	1.252	1.273	1.293	1.313	1.331	1.349	1.366	1.383	1.398	1.413
3.6	1.209	1.230	1.250	1.269	1.288	1.306	1.324	1.341	1.357	1.372
3.9	1.170	1.190	1.210	1.229	1.248	1.266	1.284	1.301	1.318	1.333
4.2	1.136	1.155	1.174	1.193	1.211	1.229	1.247	1.264	1.281	1.297
4.5	1.107	1.124	1.142	1.160	1.178	1.195	1.213	1.230	1.246	1.262
4.8	1.082	1.097	1.114	1.130	1.147	1.164	1.181	1.198	1.214	1.230
5.1	1.061	1.075	1.089	1.104	1.120	1.136	1.152	1.169	1.184	1.200
5.4	1.044	1.056	1.068	1.082	1.096	1.111	1.126	1.142	1.157	1.172
5.7	1.030	1.041	1.051	1.063	1.075	1.089	1.103	1.117	1.132	1.147
6.0	1.016	1.028	1.037	1.047	1.058	1.070	1.082	1.096	1.110	1.123
6.3	1.000	1.015	1.026	1.034	1.043	1.053	1.065	1.077	1.089	1.102
6.6	-	1.000	1.014	1.024	1.032	1.040	1.050	1.060	1.071	1.083
6.9	-	-	1.000	1.013	1.022	1.029	1.037	1.046	1.056	1.067
7.2	-	-	-	1.000	1.012	1.021	1.027	1.035	1.043	1.052
7.5	-	-	-	-	1.000	1.012	1.019	1.025	1.032	1.040
7.8	-	-	-	-	-	1.000	1.011	1.018	1.024	1.030
8.1	-	-	-	-	-	-	1.000	1.010	1.017	1.022
8.4	-	-	-	-	-	-	-	1.000	1.010	1.016
8.7	-	-	-	-	-	-	-	-	1.000	1.009
9.0	-	-	-	-	-	-	-	-	-	1.000

If you have a negative *DNS* value, use Table 2 to obtain an approximate Effective Cost of Production. To do this divide the *DNS* value by 3 and divide the *Elbow Room* value by 3 (yes 3) to obtain the corresponding values for C_{pk} and C_p to use with Table 2 and look up the Effective Cost.

Table 31: Target at boundary, rework at 100% of nominal

Elbow Room in standard deviation units

DNS	9.3	9.6	9.9	10.2	10.5	10.8	11.1	11.4	11.7	12.0
0.0	1.923	1.925	1.928	1.930	1.932	1.933	1.935	1.937	1.939	1.940
0.3	1.893	1.896	1.899	1.902	1.905	1.907	1.910	1.912	1.914	1.916
0.6	1.856	1.860	1.864	1.868	1.872	1.875	1.878	1.881	1.884	1.887
0.9	1.812	1.818	1.823	1.828	1.832	1.837	1.841	1.845	1.849	1.852
1.2	1.764	1.770	1.777	1.783	1.788	1.794	1.799	1.804	1.809	1.813
1.5	1.713	1.721	1.728	1.735	1.742	1.749	1.755	1.761	1.766	1.771
1.8	1.661	1.670	1.679	1.687	1.695	1.703	1.710	1.717	1.723	1.729
2.1	1.611	1.621	1.631	1.640	1.649	1.657	1.665	1.673	1.681	1.688
2.4	1.562	1.573	1.584	1.594	1.604	1.614	1.622	1.631	1.639	1.647
2.7	1.515	1.527	1.539	1.550	1.561	1.571	1.581	1.590	1.599	1.608
3.0	1.470	1.484	1.496	1.508	1.519	1.530	1.541	1.551	1.560	1.569
3.3	1.428	1.442	1.455	1.467	1.479	1.491	1.502	1.513	1.523	1.533
3.6	1.387	1.401	1.415	1.428	1.441	1.453	1.465	1.476	1.487	1.497
3.9	1.349	1.363	1.378	1.391	1.404	1.417	1.429	1.441	1.452	1.463
4.2	1.312	1.327	1.342	1.356	1.369	1.382	1.395	1.407	1.418	1.429
4.5	1.278	1.293	1.308	1.322	1.336	1.349	1.362	1.374	1.386	1.398
4.8	1.246	1.261	1.276	1.290	1.304	1.317	1.330	1.343	1.355	1.367
5.1	1.216	1.231	1.245	1.260	1.274	1.287	1.300	1.313	1.326	1.338
5.4	1.187	1.202	1.217	1.231	1.245	1.259	1.272	1.285	1.297	1.309
5.7	1.161	1.176	1.190	1.204	1.218	1.232	1.245	1.258	1.270	1.283
6.0	1.137	1.151	1.165	1.179	1.193	1.206	1.219	1.232	1.245	1.257
6.3	1.116	1.129	1.142	1.156	1.169	1.182	1.195	1.208	1.220	1.233
6.6	1.096	1.109	1.121	1.134	1.147	1.160	1.172	1.185	1.197	1.209
6.9	1.078	1.090	1.102	1.114	1.127	1.139	1.151	1.164	1.176	1.188
7.2	1.063	1.073	1.085	1.096	1.108	1.120	1.132	1.143	1.155	1.167
7.5	1.049	1.059	1.069	1.080	1.091	1.102	1.113	1.125	1.136	1.148
7.8	1.038	1.046	1.055	1.065	1.075	1.086	1.097	1.107	1.118	1.129
8.1	1.028	1.035	1.043	1.052	1.061	1.071	1.081	1.091	1.102	1.113
8.4	1.021	1.026	1.033	1.041	1.049	1.058	1.067	1.077	1.087	1.097
8.7	1.015	1.020	1.025	1.031	1.038	1.046	1.055	1.064	1.073	1.083
9.0	1.009	1.014	1.018	1.023	1.029	1.036	1.044	1.052	1.061	1.069
9.3	1.000	1.008	1.013	1.017	1.022	1.028	1.034	1.042	1.049	1.058
9.6	-	1.000	1.008	1.013	1.016	1.021	1.026	1.033	1.040	1.047
9.9	-	-	1.000	1.008	1.012	1.015	1.020	1.025	1.031	1.038
10.2	-	-	-	1.000	1.007	1.011	1.015	1.019	1.024	1.029
10.5	-	-	-	-	1.000	1.007	1.011	1.014	1.018	1.023
10.8	-	-	-	-	-	1.000	1.007	1.010	1.013	1.017
11.1	-	-	-	-	-	-	1.000	1.006	1.010	1.013
11.4	-	-	-	-	-	-	-	1.000	1.006	1.009
11.7	-	-	-	-	-	-	-	-	1.000	1.006

Table 31: Target at boundary, rework at 100% of nominal

	Elbow Room in std. dev. units				
DNS	***15.***	***18.***	***24.***	***36.***	***60.***
0.0	1.95	1.96	1.97	1.98	1.99
0.3	1.93	1.94	1.95	1.97	1.98
0.6	1.90	1.92	1.94	1.96	1.97
0.9	1.87	1.89	1.92	1.95	1.97
1.2	1.84	1.87	1.90	1.93	1.96
1.5	1.81	1.84	1.88	1.92	1.95
1.8	1.78	1.81	1.86	1.90	1.94
2.1	1.74	1.78	1.83	1.89	1.93
2.4	1.71	1.75	1.81	1.87	1.92
2.7	1.68	1.73	1.79	1.86	1.91
3.0	1.64	1.70	1.77	1.84	1.90
3.3	1.61	1.67	1.75	1.83	1.89
3.6	1.58	1.64	1.72	1.81	1.88
3.9	1.55	1.62	1.70	1.80	1.87
4.2	1.52	1.59	1.68	1.78	1.87
4.5	1.49	1.57	1.66	1.77	1.86
4.8	1.47	1.54	1.64	1.75	1.85
5.1	1.44	1.52	1.62	1.74	1.84
5.4	1.41	1.49	1.60	1.72	1.83
5.7	1.39	1.47	1.58	1.71	1.82
6.0	1.36	1.45	1.56	1.70	1.81
7.5	1.25	1.34	1.47	1.63	1.77
9.0	1.16	1.25	1.39	1.56	1.72
12.	1.04	1.11	1.25	1.45	1.64
15.	1.00	1.03	1.14	1.34	1.56
18.	-	1.00	1.06	1.25	1.49
24.	-	-	1.00	1.11	1.36
36.	-	-	-	1.00	1.16
60.	-	-	-	-	1.00

Table 32: The Effective Cost of Production when the target is at a boundary and nonconforming units are reworked at 80% of nominal

Elbow Room in standard deviation units

DNS	*0.3*	*0.6*	*0.9*	*1.2*	*1.5*	*1.8*	*2.1*	*2.4*	*2.7*	*3.0*
0.0	1.211	1.320	1.402	1.462	1.508	1.543	1.571	1.594	1.613	1.628
0.3	1.000	1.140	1.248	1.331	1.394	1.443	1.482	1.514	1.540	1.562
0.6	-	1.000	1.104	1.198	1.275	1.336	1.385	1.425	1.458	1.485
0.9	-	-	1.000	1.082	1.163	1.231	1.287	1.333	1.372	1.404
1.2	-	-	-	1.000	1.067	1.136	1.195	1.245	1.287	1.324
1.5	-	-	-	-	1.000	1.057	1.115	1.166	1.210	1.248
1.8	-	-	-	-	-	1.000	1.048	1.099	1.143	1.181
2.1	-	-	-	-	-	-	1.000	1.042	1.086	1.123
2.4	-	-	-	-	-	-	-	1.000	1.037	1.075
2.7	-	-	-	-	-	-	-	-	1.000	1.033
3.0	-	-	-	-	-	-	-	-	-	1.000

Elbow Room in standard deviation units

DNS	*3.3*	*3.6*	*3.9*	*4.2*	*4.5*	*4.8*	*5.1*	*5.4*	*5.7*	*6.0*
0.0	1.641	1.653	1.633	1.671	1.679	1.686	1.692	1.697	1.702	1.707
0.3	1.580	1.596	1.610	1.622	1.632	1.642	1.650	1.658	1.664	1.671
0.6	1.509	1.529	1.547	1.563	1.576	1.589	1.600	1.610	1.619	1.627
0.9	1.433	1.457	1.479	1.497	1.514	1.529	1.543	1.555	1.567	1.577
1.2	1.355	1.383	1.407	1.429	1.449	1.467	1.483	1.497	1.510	1.523
1.5	1.282	1.311	1.338	1.362	1.384	1.403	1.421	1.438	1.453	1.467
1.8	1.215	1.245	1.273	1.298	1.321	1.342	1.362	1.380	1.396	1.412
2.1	1.157	1.187	1.214	1.239	1.263	1.285	1.305	1.325	1.343	1.359
2.4	1.107	1.137	1.163	1.188	1.211	1.233	1.254	1.274	1.292	1.310
2.7	1.066	1.094	1.119	1.143	1.165	1.186	1.207	1.227	1.246	1.264
3.0	1.029	1.058	1.083	1.105	1.126	1.146	1.166	1.185	1.204	1.222
3.3	1.000	1.026	1.052	1.073	1.093	1.111	1.130	1.148	1.166	1.184
3.6	-	1.000	1.024	1.047	1.065	1.082	1.099	1.116	1.133	1.150
3.9	-	-	1.000	1.022	1.042	1.058	1.074	1.089	1.104	1.120
4.2	-	-	-	1.000	1.020	1.038	1.052	1.066	1.080	1.094
4.5	-	-	-	-	1.000	1.018	1.034	1.047	1.059	1.072
4.8	-	-	-	-	-	1.000	1.017	1.031	1.043	1.054
5.1	-	-	-	-	-	-	1.000	1.016	1.029	1.039
5.4	-	-	-	-	-	-	-	1.000	1.014	1.026
5.7	-	-	-	-	-	-	-	-	1.000	1.013
6.0	-	-	-	-	-	-	-	-	-	1.000

Table 32: Target at boundary, rework at 80% of nominal

	Elbow Room in standard deviation units									
DNS	*6.3*	*6.6*	*6.9*	*7.2*	*7.5*	*7.8*	*8.1*	*8.4*	*8.7*	*9.0*
0.0	1.711	1.715	1.718	1.721	1.724	1.727	1.730	1.732	1.734	1.736
0.3	1.676	1.682	1.686	1.691	1.695	1.699	1.702	1.706	1.709	1.712
0.6	1.634	1.641	1.648	1.653	1.659	1.664	1.669	1.673	1.677	1.681
0.9	1.586	1.595	1.603	1.610	1.617	1.623	1.629	1.635	1.640	1.645
1.2	1.534	1.544	1.554	1.563	1.571	1.579	1.586	1.593	1.599	1.605
1.5	1.480	1.492	1.503	1.513	1.523	1.532	1.541	1.549	1.556	1.563
1.8	1.426	1.440	1.453	1.464	1.475	1.486	1.495	1.505	1.513	1.521
2.1	1.375	1.390	1.403	1.416	1.429	1.440	1.451	1.461	1.471	1.480
2.4	1.326	1.342	1.357	1.371	1.384	1.397	1.408	1.419	1.430	1.440
2.7	1.281	1.298	1.313	1.328	1.342	1.355	1.368	1.380	1.391	1.402
3.0	1.240	1.256	1.272	1.288	1.302	1.316	1.329	1.342	1.354	1.365
3.3	1.202	1.218	1.235	1.250	1.265	1.279	1.293	1.306	1.319	1.331
3.6	1.167	1.184	1.200	1.215	1.231	1.245	1.259	1.273	1.285	1.298
3.9	1.136	1.152	1.168	1.183	1.199	1.213	1.227	1.241	1.254	1.267
4.2	1.109	1.124	1.139	1.154	1.169	1.184	1.198	1.211	1.225	1.237
4.5	1.085	1.099	1.114	1.128	1.142	1.156	1.170	1.184	1.197	1.210
4.8	1.065	1.078	1.091	1.104	1.118	1.131	1.145	1.158	1.171	1.184
5.1	1.049	1.060	1.071	1.083	1.096	1.109	1.122	1.135	1.148	1.160
5.4	1.035	1.045	1.055	1.065	1.077	1.089	1.101	1.113	1.126	1.138
5.7	1.024	1.032	1.041	1.050	1.060	1.071	1.082	1.094	1.106	1.117
6.0	1.013	1.022	1.030	1.038	1.046	1.056	1.066	1.077	1.088	1.099
6.3	1.000	1.012	1.021	1.027	1.035	1.043	1.052	1.061	1.071	1.082
6.6	-	1.000	1.011	1.019	1.025	1.032	1.040	1.048	1.057	1.067
6.9	-	-	1.000	1.010	1.018	1.024	1.030	1.037	1.045	1.053
7.2	-	-	-	1.000	1.010	1.017	1.022	1.028	1.034	1.042
7.5	-	-	-	-	1.000	1.009	1.016	1.020	1.026	1.032
7.8	-	-	-	-	-	1.000	1.009	1.015	1.019	1.024
8.1	-	-	-	-	-	-	1.000	1.008	1.014	1.018
8.4	-	-	-	-	-	-	-	1.000	1.008	1.013
8.7	-	-	-	-	-	-	-	-	1.000	1.007
9.0	-	-	-	-	-	-	-	-	-	1.000

If you have a negative *DNS* value, use Table 6 to obtain an approximate Effective Cost of Production. To do this divide the *DNS* value by 3 and divide the *Elbow Room* value by 3 (yes 3) to obtain the corresponding values for C_{pk} and C_p to use with Table 6 and look up the Effective Cost.

Table 32: Target at boundary, rework at 80% of nominal

	Elbow Room in standard deviation units									
DNS	***9.3***	***9.6***	***9.9***	***10.2***	***10.5***	***10.8***	***11.1***	***11.4***	***11.7***	***12.0***
0.0	1.738	1.740	1.742	1.744	1.745	1.747	1.748	1.750	1.751	1.752
0.3	1.714	1.717	1.719	1.722	1.724	1.726	1.728	1.730	1.732	1.733
0.6	1.685	1.688	1.691	1.694	1.697	1.700	1.703	1.705	1.707	1.710
0.9	1.650	1.654	1.658	1.662	1.666	1.669	1.673	1.676	1.679	1.682
1.2	1.611	1.616	1.621	1.626	1.631	1.635	1.639	1.643	1.647	1.650
1.5	1.570	1.577	1.583	1.588	1.594	1.599	1.604	1.608	1.613	1.617
1.8	1.529	1.536	1.543	1.550	1.556	1.562	1.568	1.573	1.578	1.583
2.1	1.489	1.497	1.505	1.512	1.519	1.526	1.532	1.539	1.544	1.550
2.4	1.450	1.459	1.467	1.476	1.483	1.491	1.498	1.505	1.511	1.518
2.7	1.412	1.422	1.431	1.440	1.449	1.457	1.465	1.472	1.479	1.486
3.0	1.376	1.387	1.397	1.406	1.415	1.424	1.432	1.441	1.448	1.456
3.3	1.342	1.353	1.364	1.374	1.383	1.393	1.402	1.410	1.418	1.426
3.6	1.310	1.321	1.332	1.343	1.353	1.362	1.372	1.381	1.389	1.398
3.9	1.279	1.291	1.302	1.313	1.323	1.333	1.343	1.352	1.361	1.370
4.2	1.250	1.262	1.273	1.285	1.295	1.306	1.316	1.325	1.335	1.344
4.5	1.222	1.234	1.246	1.258	1.268	1.279	1.289	1.299	1.309	1.318
4.8	1.197	1.209	1.220	1.232	1.243	1.254	1.264	1.274	1.284	1.294
5.1	1.172	1.184	1.196	1.208	1.219	1.230	1.240	1.250	1.260	1.270
5.4	1.150	1.162	1.173	1.185	1.196	1.207	1.217	1.228	1.238	1.248
5.7	1.129	1.141	1.152	1.163	1.174	1.185	1.196	1.206	1.216	1.226
6.0	1.110	1.121	1.132	1.143	1.154	1.165	1.175	1.186	1.196	1.206
6.3	1.092	1.103	1.114	1.125	1.135	1.146	1.156	1.166	1.176	1.186
6.6	1.077	1.087	1.097	1.107	1.118	1.128	1.138	1.148	1.158	1.168
6.9	1.063	1.072	1.082	1.091	1.101	1.111	1.121	1.131	1.140	1.150
7.2	1.050	1.059	1.068	1.077	1.086	1.096	1.105	1.115	1.124	1.134
7.5	1.039	1.047	1.055	1.064	1.073	1.082	1.091	1.100	1.109	1.118
7.8	1.030	1.037	1.044	1.052	1.060	1.069	1.077	1.086	1.095	1.104
8.1	1.023	1.028	1.035	1.042	1.049	1.057	1.065	1.073	1.082	1.090
8.4	1.017	1.021	1.027	1.033	1.039	1.046	1.054	1.062	1.069	1.078
8.7	1.012	1.016	1.020	1.025	1.031	1.037	1.044	1.051	1.058	1.066
9.0	1.007	1.011	1.015	1.019	1.024	1.029	1.035	1.042	1.048	1.056
9.3	1.000	1.007	1.011	1.014	1.018	1.022	1.028	1.033	1.040	1.046
9.6	-	1.000	1.006	1.010	1.013	1.017	1.021	1.026	1.032	1.038
9.9	-	-	1.000	1.006	1.010	1.012	1.016	1.020	1.025	1.030
10.2	-	-	-	1.000	1.006	1.009	1.012	1.015	1.019	1.024
10.5	-	-	-	-	1.000	1.006	1.009	1.011	1.014	1.018
10.8	-	-	-	-	-	1.000	1.005	1.008	1.011	1.014
11.1	-	-	-	-	-	-	1.000	1.005	1.008	1.010
11.4	-	-	-	-	-	-	-	1.000	1.005	1.007
11.7	-	-	-	-	-	-	-	-	1.000	1.005

Table 32: Target at boundary, rework at 80% of nominal

DNS	Elbow Room in std. dev. units				
	15.	18.	24.	36.	60.
0.0	1.76	1.77	1.77	1.78	1.79
0.3	1.74	1.75	1.76	1.78	1.79
0.6	1.72	1.74	1.75	1.77	1.78
0.9	1.70	1.72	1.74	1.76	1.77
1.2	1.67	1.69	1.72	1.75	1.77
1.5	1.65	1.67	1.70	1.73	1.76
1.8	1.62	1.65	1.69	1.72	1.75
2.1	1.59	1.63	1.67	1.71	1.75
2.4	1.57	1.60	1.65	1.70	1.74
2.7	1.54	1.58	1.63	1.69	1.73
3.0	1.52	1.56	1.61	1.67	1.72
3.3	1.49	1.54	1.60	1.66	1.71
3.6	1.47	1.51	1.58	1.65	1.71
3.9	1.44	1.49	1.56	1.64	1.70
4.2	1.42	1.47	1.55	1.62	1.69
4.5	1.40	1.45	1.53	1.61	1.68
4.8	1.37	1.43	1.51	1.60	1.68
5.1	1.35	1.41	1.50	1.59	1.67
5.4	1.33	1.39	1.48	1.58	1.66
5.7	1.31	1.38	1.47	1.57	1.66
6.0	1.29	1.36	1.45	1.56	1.65
7.5	1.20	1.27	1.38	1.50	1.61
9.0	1.13	1.20	1.31	1.45	1.58
12.	1.04	1.09	1.20	1.36	1.51
15.	1.00	1.02	1.11	1.27	1.45
18.	-	1.00	1.05	1.20	1.39
24.	-	-	1.00	1.09	1.29
36.	-	-	-	1.00	1.13
60.	-	-	-	-	1.00

Table 33: The Effective Cost of Production when the target is at a boundary and nonconforming units are reworked at 67% of nominal

Elbow Room in standard deviation units

DNS	0.3	0.6	0.9	1.2	1.5	1.8	2.1	2.4	2.7	3.0
0.0	1.176	1.267	1.336	1.388	1.423	1.453	1.476	1.495	1.511	1.524
0.3	1.000	1.117	1.206	1.276	1.329	1.370	1.402	1.429	1.450	1.468
0.6	-	1.000	1.087	1.165	1.229	1.280	1.321	1.354	1.382	1.405
0.9	-	-	1.000	1.069	1.136	1.192	1.239	1.277	1.310	1.337
1.2	-	-	-	1.000	1.056	1.113	1.163	1.204	1.240	1.270
1.5	-	-	-	-	1.000	1.047	1.096	1.139	1.175	1.207
1.8	-	-	-	-	-	1.000	1.040	1.083	1.119	1.151
2.1	-	-	-	-	-	-	1.000	1.035	1.071	1.103
2.4	-	-	-	-	-	-	-	1.000	1.031	1.062
2.7	-	-	-	-	-	-	-	-	1.000	1.027
3.0	-	-	-	-	-	-	-	-	-	1.000

Elbow Room in standard deviation units

DNS	3.3	3.6	3.9	4.2	4.5	4.8	5.1	5.4	5.7	6.0
0.0	1.535	1.544	1.553	1.560	1.566	1.572	1.577	1.581	1.586	1.589
0.3	1.484	1.497	1.508	1.518	1.527	1.535	1.542	1.548	1.554	1.559
0.6	1.424	1.441	1.456	1.469	1.481	1.491	1.500	1.508	1.516	1.523
0.9	1.361	1.381	1.399	1.415	1.429	1.441	1.453	1.463	1.472	1.481
1.2	1.296	1.319	1.340	1.358	1.374	1.389	1.402	1.414	1.426	1.436
1.5	1.235	1.260	1.282	1.302	1.320	1.336	1.351	1.365	1.378	1.389
1.8	1.179	1.205	1.227	1.248	1.268	1.285	1.302	1.317	1.331	1.343
2.1	1.131	1.156	1.179	1.200	1.219	1.238	1.255	1.271	1.286	1.300
2.4	1.089	1.114	1.136	1.156	1.176	1.194	1.212	1.228	1.244	1.258
2.7	1.055	1.078	1.100	1.119	1.138	1.155	1.173	1.189	1.205	1.220
3.0	1.024	1.049	1.069	1.088	1.105	1.122	1.138	1.154	1.170	1.185
3.3	1.000	1.022	1.043	1.061	1.077	1.093	1.108	1.124	1.139	1.154
3.6	-	1.000	1.020	1.039	1.054	1.069	1.083	1.097	1.111	1.125
3.9	-	-	1.000	1.018	1.035	1.049	1.061	1.074	1.087	1.100
4.2	-	-	-	1.000	1.017	1.032	1.044	1.055	1.066	1.078
4.5	-	-	-	-	1.000	1.015	1.029	1.039	1.050	1.060
4.8	-	-	-	-	-	1.000	1.014	1.026	1.036	1.045
5.1	-	-	-	-	-	-	1.000	1.013	1.024	1.032
5.4	-	-	-	-	-	-	-	1.000	1.012	1.022
5.7	-	-	-	-	-	-	-	-	1.000	1.011
6.0	-	-	-	-	-	-	-	-	-	1.000

Table 33: Target at boundary, rework at 67% of nominal

	Elbow Room in standard deviation units									
DNS	***6.3***	***6.6***	***6.9***	***7.2***	***7.5***	***7.8***	***8.1***	***8.4***	***8.7***	***9.0***
0.0	1.593	1.596	1.599	1.601	1.604	1.606	1.608	1.610	1.612	1.614
0.3	1.564	1.568	1.572	1.576	1.579	1.583	1.586	1.588	1.591	1.593
0.6	1.529	1.535	1.540	1.545	1.549	1.554	1.557	1.561	1.565	1.568
0.9	1.489	1.496	1.503	1.509	1.514	1.520	1.525	1.529	1.534	1.538
1.2	1.445	1.454	1.462	1.469	1.476	1.483	1.489	1.494	1.500	1.505
1.5	1.400	1.410	1.419	1.428	1.436	1.444	1.451	1.458	1.464	1.470
1.8	1.356	1.367	1.377	1.387	1.396	1.405	1.413	1.421	1.428	1.435
2.1	1.313	1.325	1.336	1.347	1.357	1.367	1.376	1.384	1.393	1.400
2.4	1.272	1.285	1.298	1.309	1.320	1.331	1.340	1.350	1.359	1.367
2.7	1.235	1.248	1.261	1.273	1.285	1.296	1.307	1.317	1.326	1.335
3.0	1.200	1.214	1.227	1.240	1.252	1.264	1.275	1.285	1.295	1.305
3.3	1.168	1.182	1.196	1.209	1.221	1.233	1.244	1.255	1.266	1.276
3.6	1.139	1.153	1.167	1.180	1.192	1.204	1.216	1.227	1.238	1.248
3.9	1.114	1.127	1.140	1.153	1.166	1.178	1.189	1.201	1.212	1.222
4.2	1.091	1.104	1.116	1.129	1.141	1.153	1.165	1.176	1.187	1.198
4.5	1.071	1.083	1.095	1.107	1.119	1.130	1.142	1.153	1.164	1.175
4.8	1.055	1.065	1.076	1.087	1.098	1.110	1.121	1.132	1.143	1.153
5.1	1.041	1.050	1.059	1.070	1.080	1.091	1.102	1.112	1.123	1.133
5.4	1.030	1.037	1.046	1.055	1.064	1.074	1.084	1.095	1.105	1.115
5.7	1.020	1.027	1.034	1.042	1.050	1.059	1.069	1.078	1.088	1.098
6.0	1.010	1.019	1.025	1.031	1.039	1.046	1.055	1.064	1.073	1.082
6.3	1.000	1.010	1.017	1.023	1.029	1.036	1.043	1.051	1.060	1.068
6.6	-	1.000	1.009	1.016	1.021	1.027	1.033	1.040	1.048	1.056
6.9	-	-	1.000	1.009	1.015	1.020	1.025	1.031	1.037	1.045
7.2	-	-	-	1.000	1.008	1.014	1.018	1.023	1.029	1.035
7.5	-	-	-	-	1.000	1.008	1.013	1.017	1.021	1.027
7.8	-	-	-	-	-	1.000	1.007	1.012	1.016	1.020
8.1	-	-	-	-	-	-	1.000	1.007	1.011	1.015
8.4	-	-	-	-	-	-	-	1.000	1.006	1.011
8.7	-	-	-	-	-	-	-	-	1.000	1.006
9.0	-	-	-	-	-	-	-	-	-	1.000

If you have a negative *DNS* value, use Table 8 to obtain an approximate Effective Cost of Production. To do this divide the *DNS* value by 3 and divide the *Elbow Room* value by 3 (yes 3) to obtain the corresponding values for C_{pk} and C_p to use with Table 8 and look up the Effective Cost.

Table 33: Target at boundary, rework at 67% of nominal

DNS	*Elbow Room in standard deviation units*									
	9.3	**9.6**	**9.9**	**10.2**	**10.5**	**10.8**	**11.1**	**11.4**	**11.7**	**12.0**
0.0	1.616	1.617	1.619	1.620	1.621	1.623	1.624	1.625	1.626	1.627
0.3	1.596	1.598	1.600	1.602	1.604	1.605	1.607	1.608	1.610	1.611
0.6	1.571	1.574	1.576	1.579	1.581	1.584	1.586	1.588	1.590	1.592
0.9	1.542	1.545	1.549	1.552	1.555	1.558	1.561	1.563	1.566	1.568
1.2	1.509	1.514	1.518	1.522	1.526	1.529	1.533	1.536	1.539	1.542
1.5	1.475	1.481	1.486	1.490	1.495	1.499	1.503	1.507	1.511	1.515
1.8	1.441	1.447	1.453	1.458	1.464	1.469	1.473	1.478	1.482	1.486
2.1	1.407	1.414	1.421	1.427	1.433	1.439	1.444	1.449	1.454	1.459
2.4	1.375	1.382	1.390	1.396	1.403	1.409	1.415	1.421	1.426	1.432
2.7	1.344	1.352	1.360	1.367	1.374	1.381	1.387	1.394	1.400	1.405
3.0	1.314	1.322	1.331	1.339	1.346	1.354	1.361	1.367	1.374	1.380
3.3	1.285	1.294	1.303	1.312	1.320	1.327	1.335	1.342	1.349	1.355
3.6	1.258	1.268	1.277	1.286	1.294	1.302	1.310	1.317	1.325	1.331
3.9	1.233	1.242	1.252	1.261	1.270	1.278	1.286	1.294	1.301	1.309
4.2	1.208	1.218	1.228	1.237	1.246	1.255	1.263	1.271	1.279	1.286
4.5	1.185	1.195	1.205	1.215	1.224	1.233	1.241	1.249	1.257	1.265
4.8	1.164	1.174	1.184	1.193	1.203	1.212	1.220	1.229	1.237	1.245
5.1	1.144	1.154	1.164	1.173	1.182	1.192	1.200	1.209	1.217	1.225
5.4	1.125	1.135	1.145	1.154	1.163	1.172	1.181	1.190	1.198	1.206
5.7	1.108	1.117	1.127	1.136	1.145	1.154	1.163	1.172	1.180	1.188
6.0	1.092	1.101	1.110	1.120	1.129	1.137	1.146	1.155	1.163	1.171
6.3	1.077	1.086	1.095	1.104	1.113	1.122	1.130	1.139	1.147	1.155
6.6	1.064	1.072	1.081	1.089	1.098	1.107	1.115	1.123	1.132	1.140
6.9	1.052	1.060	1.068	1.076	1.084	1.093	1.101	1.109	1.117	1.125
7.2	1.042	1.049	1.056	1.064	1.072	1.080	1.088	1.096	1.104	1.111
7.5	1.033	1.039	1.046	1.053	1.060	1.068	1.076	1.083	1.091	1.098
7.8	1.025	1.031	1.037	1.043	1.050	1.057	1.064	1.072	1.079	1.086
8.1	1.019	1.024	1.029	1.035	1.041	1.047	1.054	1.061	1.068	1.075
8.4	1.014	1.018	1.022	1.027	1.033	1.039	1.045	1.051	1.058	1.065
8.7	1.010	1.013	1.017	1.021	1.026	1.031	1.037	1.043	1.049	1.055
9.0	1.006	1.009	1.012	1.016	1.020	1.024	1.029	1.035	1.040	1.046
9.3	1.000	1.006	1.009	1.012	1.015	1.019	1.023	1.028	1.033	1.038
9.6	-	1.000	1.005	1.008	1.011	1.014	1.018	1.022	1.026	1.031
9.9	-	-	1.000	1.005	1.008	1.010	1.013	1.017	1.021	1.025
10.2	-	-	-	1.000	1.005	1.008	1.010	1.013	1.016	1.020
10.5	-	-	-	-	1.000	1.005	1.007	1.009	1.012	1.015
10.8	-	-	-	-	-	1.000	1.004	1.007	1.009	1.011
11.1	-	-	-	-	-	-	1.000	1.004	1.006	1.008
11.4	-	-	-	-	-	-	-	1.000	1.004	1.006
11.7	-	-	-	-	-	-	-	-	1.000	1.004

Table 33: Target at boundary, rework at 67% of nominal

	Elbow Room in std. dev. units				
DNS	***15.***	***18.***	***24.***	***36.***	***60.***
0.0	1.63	1.64	1.65	1.65	1.66
0.3	1.62	1.63	1.64	1.65	1.65
0.6	1.60	1.61	1.63	1.64	1.65
0.9	1.58	1.60	1.61	1.63	1.65
1.2	1.56	1.58	1.60	1.62	1.64
1.5	1.54	1.56	1.59	1.61	1.63
1.8	1.52	1.54	1.57	1.60	1.63
2.1	1.50	1.52	1.56	1.59	1.62
2.4	1.47	1.50	1.54	1.58	1.61
2.7	1.45	1.48	1.53	1.57	1.61
3.0	1.43	1.47	1.51	1.56	1.60
3.3	1.41	1.45	1.50	1.55	1.60
3.6	1.39	1.43	1.48	1.54	1.59
3.9	1.37	1.41	1.47	1.53	1.58
4.2	1.35	1.39	1.45	1.52	1.58
4.5	1.33	1.38	1.44	1.51	1.57
4.8	1.31	1.36	1.43	1.50	1.56
5.1	1.29	1.34	1.41	1.49	1.56
5.4	1.28	1.33	1.40	1.48	1.55
5.7	1.26	1.31	1.39	1.47	1.55
6.0	1.24	1.30	1.38	1.46	1.54
7.5	1.17	1.23	1.32	1.42	1.51
9.0	1.11	1.17	1.26	1.38	1.48
12.	1.03	1.08	1.17	1.30	1.43
15.	1.00	1.02	1.09	1.23	1.38
18.	-	1.00	1.04	1.17	1.33
24.	-	-	1.00	1.07	1.24
36.	-	-	-	1.00	1.11
60.	-	-	-	-	1.00

Table 34: The Effective Cost of Production when the target is at a boundary and nonconforming units are reworked at 50% of nominal

Elbow Room in standard deviation units

DNS	0.3	0.6	0.9	1.2	1.5	1.8	2.1	2.4	2.7	3.0
0.0	1.132	1.200	1.252	1.289	1.317	1.339	1.357	1.371	1.383	1.393
0.3	1.000	1.087	1.155	1.207	1.246	1.277	1.302	1.321	1.337	1.351
0.6	-	1.000	1.065	1.124	1.172	1.210	1.241	1.265	1.286	1.303
0.9	-	-	1.000	1.051	1.120	1.144	1.179	1.208	1.232	1.253
1.2	-	-	-	1.000	1.042	1.085	1.122	1.153	1.180	1.202
1.5	-	-	-	-	1.000	1.035	1.072	1.104	1.132	1.155
1.8	-	-	-	-	-	1.000	1.030	1.062	1.089	1.113
2.1	-	-	-	-	-	-	1.000	1.026	1.054	1.077
2.4	-	-	-	-	-	-	-	1.000	1.023	1.047
2.7	-	-	-	-	-	-	-	-	1.000	1.020
3.0	-	-	-	-	-	-	-	-	-	1.000

Elbow Room in standard deviation units

DNS	3.3	3.6	3.9	4.2	4.5	4.8	5.1	5.4	5.7	6.0
0.0	1.401	1.408	1.414	1.420	1.424	1.429	1.432	1.436	1.439	1.442
0.3	1.363	1.372	1.381	1.388	1.395	1.401	1.406	1.411	1.415	1.419
0.6	1.318	1.331	1.342	1.352	1.360	1.368	1.375	1.381	1.387	1.392
0.9	1.270	1.286	1.299	1.311	1.321	1.331	1.339	1.347	1.354	1.361
1.2	1.222	1.239	1.255	1.268	1.281	1.292	1.302	1.311	1.319	1.327
1.5	1.176	1.195	1.211	1.226	1.240	1.252	1.263	1.274	1.283	1.292
1.8	1.134	1.153	1.171	1.186	1.201	1.214	1.226	1.237	1.248	1.257
2.1	1.098	1.117	1.134	1.150	1.164	1.178	1.191	1.203	1.214	1.225
2.4	1.067	1.088	1.102	1.117	1.132	1.146	1.159	1.171	1.183	1.194
2.7	1.041	1.059	1.075	1.089	1.103	1.117	1.129	1.142	1.154	1.165
3.0	1.018	1.037	1.052	1.066	1.079	1.091	1.104	1.116	1.127	1.139
3.3	1.000	1.016	1.033	1.046	1.058	1.070	1.081	1.093	1.104	1.115
3.6	-	1.000	1.015	1.029	1.041	1.051	1.062	1.073	1.083	1.094
3.9	-	-	1.000	1.014	1.026	1.036	1.046	1.055	1.065	1.075
4.2	-	-	-	1.000	1.012	1.024	1.033	1.041	1.050	1.059
4.5	-	-	-	-	1.000	1.011	1.022	1.029	1.037	1.045
4.8	-	-	-	-	-	1.000	1.011	1.020	1.027	1.034
5.1	-	-	-	-	-	-	1.000	1.010	1.018	1.024
5.4	-	-	-	-	-	-	-	1.000	1.009	1.016
5.7	-	-	-	-	-	-	-	-	1.000	1.008
6.0	-	-	-	-	-	-	-	-	-	1.000

Table 34: Target at boundary, rework at 50% of nominal

	Elbow Room in standard deviation units									
DNS	*6.3*	*6.6*	*6.9*	*7.2*	*7.5*	*7.8*	*8.1*	*8.4*	*8.7*	*9.0*
0.0	1.444	1.447	1.449	1.451	1.453	1.454	1.456	1.458	1.459	1.460
0.3	1.423	1.426	1.429	1.432	1.434	1.437	1.439	1.441	1.443	1.445
0.6	1.396	1.401	1.405	1.408	1.412	1.415	1.418	1.421	1.423	1.426
0.9	1.366	1.372	1.377	1.381	1.386	1.390	1.393	1.397	1.400	1.403
1.2	1.334	1.340	1.346	1.352	1.357	1.362	1.366	1.371	1.375	1.378
1.5	1.300	1.307	1.314	1.321	1.327	1.333	1.338	1.343	1.348	1.352
1.8	1.267	1.275	1.283	1.290	1.297	1.304	1.310	1.315	1.321	1.326
2.1	1.234	1.244	1.252	1.260	1.268	1.275	1.282	1.288	1.294	1.300
2.4	1.204	1.214	1.223	1.232	1.240	1.248	1.255	1.262	1.269	1.275
2.7	1.176	1.186	1.196	1.205	1.214	1.222	1.230	1.237	1.244	1.251
3.0	1.150	1.160	1.170	1.180	1.189	1.198	1.206	1.214	1.221	1.228
3.3	1.126	1.136	1.147	1.156	1.166	1.175	1.183	1.191	1.199	1.207
3.6	1.104	1.115	1.125	1.135	1.144	1.153	1.162	1.170	1.178	1.186
3.9	1.085	1.095	1.105	1.115	1.124	1.133	1.142	1.151	1.159	1.167
4.2	1.068	1.078	1.087	1.096	1.106	1.115	1.124	1.132	1.140	1.148
4.5	1.053	1.062	1.071	1.080	1.089	1.098	1.106	1.115	1.123	1.131
4.8	1.041	1.049	1.057	1.065	1.074	1.082	1.091	1.099	1.107	1.115
5.1	1.031	1.037	1.045	1.052	1.060	1.068	1.076	1.084	1.092	1.100
5.4	1.022	1.028	1.034	1.041	1.048	1.056	1.063	1.071	1.079	1.086
5.7	1.015	1.020	1.026	1.031	1.038	1.044	1.052	1.059	1.066	1.073
6.0	1.008	1.015	1.019	1.023	1.029	1.035	1.041	1.048	1.055	1.062
6.3	1.000	1.007	1.013	1.017	1.022	1.027	1.032	1.038	1.045	1.051
6.6	-	1.000	1.007	1.012	1.016	1.020	1.025	1.030	1.036	1.042
6.9	-	-	1.000	1.006	1.011	1.015	1.019	1.023	1.028	1.033
7.2	-	-	-	1.000	1.006	1.010	1.014	1.017	1.021	1.026
7.5	-	-	-	-	1.000	1.006	1.010	1.013	1.016	1.020
7.8	-	-	-	-	-	1.000	1.005	1.009	1.012	1.015
8.1	-	-	-	-	-	-	1.000	1.005	1.009	1.011
8.4	-	-	-	-	-	-	-	1.000	1.005	1.008
8.7	-	-	-	-	-	-	-	-	1.000	1.005
9.0	-	-	-	-	-	-	-	-	-	1.000

If you have a negative *DNS* value, use Table 10 to obtain an approximate Effective Cost of Production. To do this divide the *DNS* value by 3 and divide the *Elbow Room* value by 3 (yes 3) to obtain the corresponding values for C_{pk} and C_p to use with Table 10 and look up the Effective Cost.

Table 34: Target at boundary, rework at 50% of nominal

Elbow Room in standard deviation units

DNS	9.3	9.6	9.9	10.2	10.5	10.8	11.1	11.4	11.7	12.0
0.0	1.462	1.463	1.464	1.465	1.466	1.467	1.468	1.468	1.469	1.470
0.3	1.446	1.448	1.450	1.451	1.452	1.454	1.455	1.456	1.457	1.458
0.6	1.428	1.430	1.432	1.434	1.436	1.437	1.439	1.441	1.442	1.444
0.9	1.406	1.409	1.411	1.414	1.416	1.418	1.420	1.422	1.424	1.426
1.2	1.382	1.385	1.388	1.391	1.394	1.397	1.399	1.402	1.404	1.406
1.5	1.356	1.360	1.364	1.368	1.371	1.374	1.377	1.380	1.383	1.386
1.8	1.331	1.335	1.340	1.344	1.348	1.351	1.355	1.358	1.362	1.365
2.1	1.305	1.311	1.315	1.320	1.325	1.329	1.333	1.337	1.340	1.344
2.4	1.281	1.287	1.292	1.297	1.302	1.307	1.311	1.315	1.320	1.323
2.7	1.258	1.264	1.270	1.275	1.280	1.286	1.290	1.295	1.300	1.304
3.0	1.235	1.242	1.248	1.254	1.260	1.265	1.270	1.275	1.280	1.285
3.3	1.214	1.221	1.227	1.234	1.240	1.245	1.251	1.256	1.261	1.266
3.6	1.194	1.201	1.208	1.214	1.220	1.227	1.232	1.238	1.243	1.248
3.9	1.174	1.182	1.189	1.196	1.202	1.208	1.214	1.220	1.226	1.231
4.2	1.156	1.164	1.171	1.178	1.185	1.191	1.197	1.203	1.209	1.215
4.5	1.139	1.147	1.154	1.161	1.168	1.174	1.181	1.187	1.193	1.199
4.8	1.123	1.130	1.138	1.145	1.152	1.159	1.165	1.171	1.178	1.183
5.1	1.108	1.115	1.123	1.130	1.137	1.144	1.150	1.157	1.163	1.169
5.4	1.094	1.101	1.108	1.116	1.122	1.129	1.136	1.142	1.149	1.155
5.7	1.081	1.088	1.095	1.102	1.109	1.116	1.122	1.129	1.135	1.141
6.0	1.069	1.076	1.083	1.090	1.096	1.103	1.110	1.116	1.122	1.128
6.3	1.058	1.065	1.071	1.078	1.085	1.091	1.098	1.104	1.110	1.116
6.6	1.048	1.054	1.061	1.067	1.074	1.080	1.086	1.092	1.099	1.105
6.9	1.039	1.045	1.051	1.057	1.063	1.069	1.076	1.082	1.088	1.094
7.2	1.031	1.037	1.042	1.048	1.054	1.060	1.066	1.072	1.078	1.083
7.5	1.025	1.029	1.034	1.040	1.045	1.051	1.057	1.062	1.068	1.074
7.8	1.019	1.023	1.028	1.032	1.038	1.043	1.048	1.054	1.059	1.065
8.1	1.014	1.018	1.022	1.026	1.031	1.036	1.041	1.046	1.051	1.056
8.4	1.010	1.013	1.017	1.020	1.025	1.029	1.034	1.038	1.043	1.048
8.7	1.008	1.010	1.012	1.016	1.019	1.023	1.027	1.032	1.037	1.041
9.0	1.004	1.007	1.009	1.012	1.015	1.018	1.022	1.026	1.030	1.035
9.3	1.000	1.004	1.007	1.009	1.011	1.014	1.017	1.021	1.025	1.029
9.6	-	1.000	1.004	1.006	1.008	1.010	1.013	1.016	1.020	1.023
9.9	-	-	1.000	1.004	1.006	1.008	1.010	1.013	1.015	1.019
10.2	-	-	-	1.000	1.004	1.005	1.007	1.009	1.012	1.015
10.5	-	-	-	-	1.000	1.003	1.005	1.007	1.009	1.011
10.8	-	-	-	-	-	1.000	1.003	1.005	1.007	1.008
11.1	-	-	-	-	-	-	1.000	1.003	1.005	1.006
11.4	-	-	-	-	-	-	-	1.000	1.003	1.005
11.7	-	-	-	-	-	-	-	-	1.000	1.003

Table 34: Target at boundary, rework at 50% of nominal

	Elbow Room in std. dev. units				
DNS	***15.***	***18.***	***24.***	***36.***	***60.***
0.0	1.48	1.48	1.48	1.49	1.49
0.3	1.46	1.47	1.48	1.49	1.49
0.6	1.45	1.46	1.47	1.48	1.49
0.9	1.44	1.45	1.46	1.47	1.48
1.2	1.42	1.43	1.45	1.47	1.48
1.5	1.41	1.42	1.44	1.46	1.48
1.8	1.39	1.41	1.43	1.45	1.47
2.1	1.37	1.39	1.42	1.44	1.47
2.4	1.35	1.38	1.41	1.44	1.46
2.7	1.34	1.36	1.39	1.43	1.46
3.0	1.32	1.35	1.38	1.42	1.45
3.3	1.31	1.34	1.37	1.41	1.45
3.6	1.29	1.32	1.36	1.41	1.44
3.9	1.28	1.31	1.35	1.40	1.44
4.2	1.26	1.30	1.34	1.39	1.43
4.5	1.25	1.28	1.33	1.38	1.43
4.8	1.23	1.27	1.32	1.38	1.42
5.1	1.22	1.26	1.31	1.37	1.42
5.4	1.21	1.25	1.30	1.36	1.41
5.7	1.19	1.24	1.29	1.35	1.41
6.0	1.18	1.22	1.28	1.35	1.41
7.5	1.13	1.17	1.24	1.31	1.38
9.0	1.28	1.13	1.20	1.28	1.36
12.	1.02	1.06	1.13	1.22	1.32
15.	1.00	1.02	1.07	1.17	1.28
18.	-	1.00	1.03	1.13	1.25
24.	-	-	1.00	1.06	1.18
36.	-	-	-	1.00	1.08
60.	-	-	-	-	1.00

Table 35: The Effective Cost of Production when the target is at a boundary and nonconforming units are reworked at 33% of nominal

	Elbow Room in standard deviation units									
DNS	***0.3***	***0.6***	***0.9***	***1.2***	***1.5***	***1.8***	***2.1***	***2.4***	***2.7***	***3.0***
0.0	1.088	1.133	1.168	1.192	1.211	1.226	1.238	1.247	1.255	1.261
0.3	1.000	1.058	1.103	1.138	1.164	1.185	1.201	1.214	1.225	1.234
0.6	-	1.000	1.043	1.083	1.114	1.140	1.160	1.177	1.191	1.202
0.9	-	-	1.000	1.034	1.068	1.096	1.119	1.139	1.155	1.168
1.2	-	-	-	1.000	1.028	1.057	1.081	1.102	1.120	1.135
1.5	-	-	-	-	1.000	1.024	1.048	1.069	1.088	1.103
1.8	-	-	-	-	-	1.000	1.020	1.041	1.059	1.075
2.1	-	-	-	-	-	-	1.000	1.018	1.036	1.051
2.4	-	-	-	-	-	-	-	1.000	1.015	1.031
2.7	-	-	-	-	-	-	-	-	1.000	1.014
3.0	-	-	-	-	-	-	-	-	-	1.000

	Elbow Room in standard deviation units									
DNS	***3.3***	***3.6***	***3.9***	***4.2***	***4.5***	***4.8***	***5.1***	***5.4***	***5.7***	***6.0***
0.0	1.267	1.272	1.276	1.279	1.283	1.285	1.288	1.290	1.292	1.294
0.3	1.241	1.248	1.254	1.259	1.263	1.267	1.271	1.274	1.277	1.279
0.6	1.212	1.220	1.228	1.234	1.240	1.245	1.250	1.254	1.258	1.261
0.9	1.180	1.190	1.199	1.207	1.214	1.220	1.226	1.231	1.236	1.240
1.2	1.148	1.159	1.170	1.179	1.187	1.194	1.201	1.207	1.212	1.218
1.5	1.117	1.130	1.141	1.151	1.160	1.168	1.175	1.182	1.189	1.194
1.8	1.090	1.102	1.114	1.124	1.134	1.142	1.151	1.158	1.165	1.171
2.1	1.065	1.078	1.089	1.100	1.109	1.119	1.127	1.135	1.143	1.150
2.4	1.045	1.057	1.068	1.078	1.088	1.097	1.106	1.114	1.122	1.129
2.7	1.027	1.039	1.050	1.059	1.069	1.078	1.086	1.094	1.102	1.110
3.0	1.012	1.024	1.034	1.044	1.052	1.061	1.069	1.077	1.085	1.092
3.3	1.000	1.011	1.022	1.030	1.039	1.046	1.054	1.062	1.069	1.077
3.6	-	1.000	1.010	1.019	1.027	1.034	1.041	1.048	1.055	1.063
3.9	-	-	1.000	1.009	1.017	1.024	1.031	1.037	1.043	1.050
4.2	-	-	-	1.000	1.008	1.016	1.022	1.027	1.033	1.039
4.5	-	-	-	-	1.000	1.008	1.014	1.020	1.025	1.030
4.8	-	-	-	-	-	1.000	1.007	1.013	1.018	1.022
5.1	-	-	-	-	-	-	1.000	1.006	1.012	1.016
5.4	-	-	-	-	-	-	-	1.000	1.006	1.011
5.7	-	-	-	-	-	-	-	-	1.000	1.006
6.0	-	-	-	-	-	-	-	-	-	1.000

Table 35: Target at boundary, rework at 33% of nominal

	Elbow Room in standard deviation units									
DNS	*6.3*	*6.6*	*6.9*	*7.2*	*7.5*	*7.8*	*8.1*	*8.4*	*8.7*	*9.0*
0.0	1.296	1.298	1.299	1.300	1.302	1.303	1.304	1.305	1.306	1.307
0.3	1.282	1.284	1.286	1.288	1.289	1.291	1.292	1.294	1.295	1.296
0.6	1.264	1.267	1.270	1.272	1.274	1.276	1.278	1.280	1.282	1.283
0.9	1.244	1.248	1.251	1.254	1.257	1.259	1.262	1.264	1.266	1.269
1.2	1.222	1.227	1.231	1.234	1.238	1.241	1.244	1.247	1.249	1.252
1.5	1.200	1.205	1.209	1.214	1.218	1.222	1.225	1.228	1.232	1.235
1.8	1.178	1.183	1.188	1.193	1.198	1.202	1.206	1.210	1.214	1.217
2.1	1.156	1.162	1.168	1.173	1.178	1.183	1.188	1.192	1.196	1.200
2.4	1.136	1.142	1.149	1.154	1.160	1.165	1.170	1.175	1.179	1.183
2.7	1.117	1.124	1.130	1.136	1.142	1.148	1.153	1.158	1.163	1.167
3.0	1.100	1.107	1.113	1.120	1.126	1.132	1.137	1.142	1.147	1.152
3.3	1.084	1.091	1.098	1.104	1.110	1.116	1.122	1.127	1.133	1.138
3.6	1.070	1.076	1.083	1.090	1.096	1.102	1.108	1.113	1.119	1.124
3.9	1.057	1.063	1.070	1.076	1.083	1.089	1.095	1.100	1.106	1.111
4.2	1.045	1.052	1.058	1.064	1.070	1.076	1.082	1.088	1.093	1.099
4.5	1.036	1.041	1.047	1.053	1.059	1.065	1.071	1.077	1.082	1.087
4.8	1.027	1.032	1.038	1.043	1.049	1.055	1.060	1.066	1.071	1.077
5.1	1.020	1.025	1.030	1.035	1.040	1.045	1.051	1.056	1.061	1.067
5.4	1.015	1.019	1.023	1.027	1.032	1.037	1.042	1.047	1.052	1.057
5.7	1.010	1.014	1.017	1.021	1.025	1.030	1.034	1.039	1.044	1.049
6.0	1.005	1.009	1.012	1.016	1.019	1.023	1.027	1.032	1.036	1.041
6.3	1.000	1.005	1.009	1.011	1.014	1.018	1.022	1.026	1.030	1.034
6.6	-	1.000	1.005	1.008	1.011	1.013	1.016	1.020	1.024	1.028
6.9	-	-	1.000	1.004	1.007	1.010	1.012	1.015	1.019	1.022
7.2	-	-	-	1.000	1.004	1.007	1.009	1.012	1.014	1.017
7.5	-	-	-	-	1.000	1.004	1.006	1.008	1.011	1.013
7.8	-	-	-	-	-	1.000	1.004	1.006	1.008	1.010
8.1	-	-	-	-	-	-	1.000	1.003	1.006	1.007
8.4	-	-	-	-	-	-	-	1.000	1.003	1.005
8.7	-	-	-	-	-	-	-	-	1.000	1.003
9.0	-	-	-	-	-	-	-	-	-	1.000

If you have a negative *DNS* value, use Table 12 to obtain an approximate Effective Cost of Production. To do this divide the *DNS* value by 3 and divide the *Elbow Room* value by 3 (yes 3) to obtain the corresponding values for C_{pk} and C_p to use with Table 12 and look up the Effective Cost.

Table 35: Target at boundary, rework at 33% of nominal

Elbow Room in standard deviation units

DNS	9.3	9.6	9.9	10.2	10.5	10.8	11.1	11.4	11.7	12.0
0.0	1.307	1.308	1.309	1.310	1.310	1.311	1.311	1.312	1.313	1.313
0.3	1.297	1.298	1.299	1.300	1.301	1.302	1.303	1.304	1.304	1.305
0.6	1.285	1.286	1.288	1.289	1.290	1.291	1.292	1.293	1.294	1.295
0.9	1.270	1.272	1.274	1.276	1.277	1.279	1.280	1.281	1.283	1.284
1.2	1.254	1.257	1.259	1.261	1.263	1.264	1.266	1.268	1.269	1.271
1.5	1.237	1.240	1.242	1.245	1.247	1.249	1.251	1.253	1.255	1.257
1.8	1.220	1.223	1.226	1.229	1.232	1.234	1.236	1.239	1.241	1.243
2.1	1.203	1.207	1.210	1.213	1.216	1.219	1.222	1.224	1.227	1.229
2.4	1.187	1.191	1.195	1.198	1.201	1.204	1.207	1.210	1.213	1.215
2.7	1.172	1.176	1.180	1.183	1.187	1.190	1.193	1.197	1.199	1.202
3.0	1.157	1.161	1.165	1.169	1.173	1.177	1.180	1.183	1.187	1.190
3.3	1.142	1.147	1.151	1.156	1.160	1.163	1.167	1.171	1.174	1.177
3.6	1.129	1.134	1.138	1.143	1.147	1.151	1.155	1.158	1.162	1.165
3.9	1.116	1.121	1.126	1.130	1.135	1.139	1.143	1.147	1.150	1.154
4.2	1.104	1.109	1.114	1.118	1.123	1.127	1.131	1.135	1.139	1.143
4.5	1.093	1.098	1.102	1.107	1.112	1.116	1.120	1.125	1.129	1.132
4.8	1.082	1.087	1.092	1.097	1.101	1.106	1.110	1.114	1.118	1.122
5.1	1.072	1.077	1.082	1.086	1.091	1.096	1.100	1.104	1.108	1.112
5.4	1.062	1.067	1.072	1.077	1.082	1.086	1.091	1.095	1.099	1.103
5.7	1.054	1.059	1.063	1.068	1.073	1.077	1.082	1.086	1.090	1.094
6.0	1.046	1.050	1.055	1.060	1.064	1.069	1.073	1.077	1.081	1.086
6.3	1.039	1.043	1.047	1.052	1.056	1.061	1.065	1.069	1.073	1.077
6.6	1.032	1.036	1.040	1.045	1.049	1.053	1.057	1.062	1.066	1.070
6.9	1.026	1.030	1.034	1.038	1.042	1.046	1.050	1.054	1.058	1.062
7.2	1.021	1.024	1.028	1.032	1.036	1.040	1.044	1.048	1.052	1.056
7.5	1.016	1.020	1.023	1.027	1.030	1.034	1.038	1.042	1.045	1.049
7.8	1.013	1.015	1.018	1.022	1.025	1.029	1.032	1.036	1.039	1.043
8.1	1.009	1.012	1.014	1.017	1.020	1.024	1.027	1.030	1.034	1.037
8.4	1.007	1.009	1.011	1.014	1.016	1.019	1.022	1.026	1.029	1.032
8.7	1.005	1.007	1.008	1.010	1.013	1.015	1.018	1.021	1.024	1.027
9.0	1.003	1.005	1.006	1.008	1.010	1.012	1.015	1.017	1.020	1.023
9.3	1.000	1.003	1.004	1.006	1.007	1.009	1.011	1.014	1.016	1.019
9.6	-	1.000	1.003	1.004	1.005	1.007	1.009	1.011	1.013	1.016
9.9	-	-	1.000	1.003	1.004	1.005	1.007	1.008	1.010	1.013
10.2	-	-	-	1.000	1.002	1.004	1.005	1.006	1.008	1.010
10.5	-	-	-	-	1.000	1.002	1.004	1.005	1.006	1.008
10.8	-	-	-	-	-	1.000	1.002	1.003	1.004	1.006
11.1	-	-	-	-	-	-	1.000	1.002	1.003	1.004
11.4	-	-	-	-	-	-	-	1.000	1.002	1.003
11.7	-	-	-	-	-	-	-	-	1.000	1.002

Table 35: Target at boundary, rework at 33% of nominal

DNS	*Elbow Room in std. dev. units* 15.	18.	24.	36.	60.
0.0	1.32	1.32	1.32	1.33	1.33
0.3	1.31	1.31	1.32	1.32	1.33
0.6	1.30	1.31	1.31	1.32	1.33
0.9	1.29	1.30	1.31	1.32	1.32
1.2	1.28	1.29	1.30	1.31	1.32
1.5	1.27	1.28	1.29	1.31	1.32
1.8	1.26	1.27	1.29	1.30	1.31
2.1	1.25	1.26	1.28	1.30	1.31
2.4	1.24	1.25	1.27	1.29	1.31
2.7	1.23	1.24	1.26	1.29	1.30
3.0	1.21	1.23	1.26	1.28	1.30
3.3	1.20	1.22	1.25	1.28	1.30
3.6	1.19	1.21	1.24	1.27	1.29
3.9	1.18	1.21	1.23	1.27	1.29
4.2	1.17	1.20	1.23	1.26	1.29
4.5	1.16	1.19	1.22	1.26	1.29
4.8	1.16	1.18	1.21	1.25	1.28
5.1	1.15	1.17	1.21	1.25	1.28
5.4	1.14	1.16	1.20	1.24	1.28
5.7	1.13	1.16	1.19	1.24	1.27
6.0	1.12	1.15	1.19	1.23	1.27
7.5	1.08	1.11	1.16	1.21	1.26
9.0	1.05	1.08	1.13	1.16	1.24
12.	1.01	1.04	1.08	1.15	1.21
15.	1.00	1.01	1.05	1.11	1.19
18.	-	1.00	1.02	1.08	1.16
24.	-	-	1.00	1.04	1.12
36.	-	-	-	1.00	1.05
60.	-	-	-	-	1.00

Table 36: The Effective Cost of Production when the target is at a boundary and nonconforming units are reworked at 20% of nominal

Elbow Room in standard deviation units

DNS	*0.3*	*0.6*	*0.9*	*1.2*	*1.5*	*1.8*	*2.1*	*2.4*	*2.7*	*3.0*
0.0	1.053	1.080	1.101	1.115	1.127	1.136	1.143	1.148	1.153	1.157
0.3	1.000	1.035	1.062	1.083	1.099	1.111	1.121	1.129	1.135	1.140
0.6	-	1.000	1.026	1.050	1.069	1.084	1.096	1.106	1.114	1.121
0.9	-	-	1.000	1.021	1.041	1.058	1.072	1.083	1.093	1.101
1.2	-	-	-	1.000	1.017	1.034	1.049	1.061	1.072	1.081
1.5	-	-	-	-	1.000	1.014	1.029	1.042	1.053	1.062
1.8	-	-	-	-	-	1.000	1.012	1.025	1.036	1.045
2.1	-	-	-	-	-	-	1.000	1.011	1.021	1.031
2.4	-	-	-	-	-	-	-	1.000	1.009	1.019
2.7	-	-	-	-	-	-	-	-	1.000	1.008
3.0	-	-	-	-	-	-	-	-	-	1.000

Elbow Room in standard deviation units

DNS	*3.3*	*3.6*	*3.9*	*4.2*	*4.5*	*4.8*	*5.1*	*5.4*	*5.7*	*6.0*
0.0	1.160	1.163	1.166	1.168	1.170	1.171	1.173	1.174	1.176	1.177
0.3	1.145	1.149	1.152	1.155	1.158	1.160	1.162	1.164	1.166	1.168
0.6	1.127	1.132	1.137	1.141	1.144	1.147	1.150	1.152	1.155	1.157
0.9	1.108	1.114	1.120	1.124	1.129	1.132	1.136	1.139	1.142	1.144
1.2	1.089	1.096	1.102	1.107	1.112	1.117	1.121	1.124	1.128	1.131
1.5	1.070	1.078	1.084	1.090	1.096	1.101	1.105	1.109	1.113	1.117
1.8	1.054	1.061	1.068	1.074	1.080	1.086	1.090	1.095	1.099	1.103
2.1	1.039	1.047	1.054	1.060	1.066	1.071	1.076	1.081	1.086	1.090
2.4	1.027	1.034	1.041	1.047	1.053	1.058	1.063	1.068	1.073	1.077
2.7	1.016	1.023	1.030	1.036	1.041	1.047	1.052	1.057	1.061	1.066
3.0	1.007	1.015	13.021	1.026	1.031	1.036	1.041	1.046	1.051	1.056
3.3	1.000	1.007	1.013	1.018	1.023	1.028	1.032	1.037	1.042	1.046
3.6	-	1.000	1.006	1.012	1.016	1.021	1.025	1.029	1.033	1.038
3.9	-	-	1.000	1.005	1.010	1.015	1.018	1.022	1.026	1.030
4.2	-	-	-	1.000	1.005	1.009	1.013	1.016	1.020	1.024
4.5	-	-	-	-	1.000	1.005	1.009	1.012	1.015	1.018
4.8	-	-	-	-	-	1.000	1.004	1.008	1.011	1.013
5.1	-	-	-	-	-	-	1.000	1.004	1.007	1.010
5.4	-	-	-	-	-	-	-	1.000	1.004	1.007
5.7	-	-	-	-	-	-	-	-	1.000	1.003
6.0	-	-	-	-	-	-	-	-	-	1.000

Table 36: Target at boundary, rework at 20% of nominal

DNS	Elbow Room in standard deviation units									
	6.3	6.6	6.9	7.2	7.5	7.8	8.1	8.4	8.7	9.0
0.0	1.178	1.179	1.180	1.180	1.181	1.182	1.182	1.183	1.184	1.184
0.3	1.169	1.170	1.172	1.173	1.174	1.175	1.176	1.176	1.177	1.178
0.6	1.159	1.160	1.162	1.163	1.165	1.166	1.167	1.168	1.169	1.170
0.9	1.147	1.149	1.151	1.153	1.154	1.156	1.157	1.159	1.160	1.161
1.2	1.133	1.136	1.138	1.141	1.143	1.145	1.147	1.148	1.150	1.151
1.5	1.120	1.123	1.126	1.128	1.131	1.133	1.135	1.137	1.139	1.141
1.8	1.107	1.110	1.113	1.116	1.119	1.121	1.124	1.126	1.128	1.130
2.1	1.094	1.097	1.101	1.104	1.107	1.110	1.113	1.115	1.118	1.120
2.4	1.082	1.086	1.089	1.093	1.096	1.099	1.102	1.105	1.108	1.110
2.7	1.070	1.074	1.078	1.082	1.085	1.089	1.092	1.095	1.098	1.100
3.0	1.060	1.064	1.068	1.072	1.076	1.079	1.082	1.085	1.088	1.091
3.3	1.050	1.055	1.059	1.063	1.066	1.070	1.073	1.077	1.080	1.083
3.6	1.042	1.046	1.050	1.054	1.058	1.061	1.065	1.068	1.071	1.074
3.9	1.034	1.038	1.042	1.046	1.050	1.053	1.057	1.060	1.064	1.067
4.2	1.027	1.031	1.035	1.039	1.042	1.046	1.049	1.053	1.056	1.059
4.5	1.021	1.025	1.028	1.032	1.036	1.039	1.043	1.046	1.049	1.052
4.8	1.016	1.019	1.023	1.026	1.029	1.033	1.036	1.040	1.043	1.046
5.1	1.012	1.015	1.018	1.021	1.024	1.027	1.030	1.034	1.037	1.040
5.4	1.009	1.011	1.014	1.016	1.019	1.022	1.025	1.028	1.031	1.034
5.7	1.006	1.008	1.010	1.013	1.015	1.018	1.021	1.023	1.026	1.029
6.0	1.003	1.006	1.007	1.009	1.012	1.014	1.016	1.019	1.022	1.025
6.3	1.000	1.003	1.005	1.007	1.009	1.011	1.013	1.015	1.018	1.020
6.6	-	1.000	1.003	1.005	1.006	1.008	1.010	1.012	1.014	1.017
6.9	-	-	1.000	1.003	1.004	1.006	1.007	1.009	1.011	1.013
7.2	-	-	-	1.000	1.002	1.004	1.005	1.007	1.009	1.010
7.5	-	-	-	-	1.000	1.002	1.004	1.005	1.006	1.008
7.8	-	-	-	-	-	1.000	1.002	1.004	1.005	1.006
8.1	-	-	-	-	-	-	1.000	1.002	1.003	1.004
8.4	-	-	-	-	-	-	-	1.000	1.002	1.003
8.7	-	-	-	-	-	-	-	-	1.000	1.002
9.0	-	-	-	-	-	-	-	-	-	1.000

If you have a negative *DNS* value, use Table 14 to obtain an approximate Effective Cost of Production. To do this divide the *DNS* value by 3 and divide the *Elbow Room* value by 3 (yes 3) to obtain the corresponding values for C_{pk} and C_p to use with Table 14 and look up the Effective Cost.

Table 36: Target at boundary, rework at 20% of nominal

Elbow Room in standard deviation units

DNS	*9.3*	*9.6*	*9.9*	*10.2*	*10.5*	*10.8*	*11.1*	*11.4*	*11.7*	*12.0*
0.0	1.185	1.185	1.186	1.186	1.186	1.187	1.187	1.187	1.188	1.188
0.3	1.179	1.179	1.180	1.180	1.181	1.181	1.182	1.182	1.183	1.183
0.6	1.171	1.172	1.173	1.174	1.174	1.175	1.176	1.176	1.177	1.177
0.9	1.162	1.164	1.165	1.166	1.166	1.167	1.168	1.169	1.170	1.170
1.2	1.153	1.154	1.155	1.157	1.158	1.159	1.160	1.161	1.162	1.163
1.5	1.143	1.144	1.146	1.147	1.148	1.150	1.151	1.152	1.153	1.154
1.8	1.132	1.134	1.136	1.137	1.139	1.141	1.142	1.143	1.145	1.146
2.1	1.122	1.124	1.126	1.128	1.130	1.131	1.133	1.135	1.136	1.138
2.4	1.112	1.115	1.117	1.119	1.121	1.123	1.124	1.126	1.128	1.129
2.7	1.103	1.105	1.108	1.110	1.112	1.114	1.116	1.118	1.120	1.122
3.0	1.094	1.097	1.099	1.102	1.104	1.106	1.108	1.110	1.112	1.114
3.3	1.086	1.088	1.091	1.093	1.096	1.098	1.100	1.103	1.105	1.107
3.6	1.077	1.080	1.083	1.086	1.088	1.091	1.093	1.095	1.097	1.099
3.9	1.070	1.073	1.076	1.078	1.081	1.083	1.086	1.088	1.090	1.093
4.2	1.062	1.065	1.068	1.071	1.074	1.076	1.079	1.081	1.084	1.086
4.5	1.056	1.059	1.062	1.064	1.067	1.070	1.072	1.075	1.077	1.080
4.8	1.049	1.052	1.055	1.058	1.061	1.063	1.066	1.069	1.071	1.073
5.1	1.043	1.046	1.049	1.052	1.055	1.057	1.060	1.063	1.065	1.068
5.4	1.037	1.040	1.043	1.046	1.049	1.052	1.054	1.057	1.059	1.062
5.7	1.032	1.035	1.038	1.041	1.044	1.046	1.049	1.052	1.054	1.057
6.0	1.027	1.030	1.033	1.036	1.039	1.041	1.044	1.046	1.049	1.051
6.3	1.023	1.026	1.028	1.031	1.034	1.036	1.039	1.042	1.044	1.047
6.6	1.019	1.022	1.024	1.027	1.029	1.032	1.034	1.037	1.039	1.042
6.9	1.016	1.018	1.020	1.023	1.025	1.028	1.030	1.033	1.035	1.038
7.2	1.013	1.015	1.017	1.019	1.022	1.024	1.026	1.029	1.031	1.033
7.5	1.010	1.012	1.014	1.016	1.018	1.020	1.023	1.025	1.027	1.030
7.8	1.008	1.009	1.011	1.013	1.015	1.017	1.019	1.021	1.024	1.026
8.1	1.006	1.007	1.009	1.010	1.012	1.014	1.016	1.018	1.020	1.023
8.4	1.004	1.005	1.007	1.008	1.010	1.012	1.013	1.015	1.017	1.019
8.7	1.003	1.004	1.005	1.006	1.008	1.009	1.011	1.013	1.015	1.017
9.0	1.002	1.003	1.004	1.005	1.006	1.007	1.009	1.010	1.012	1.014
9.3	1.000	1.002	1.003	1.003	1.004	1.006	1.007	1.008	1.010	1.012
9.6	-	1.000	1.002	1.003	1.003	1.004	1.005	1.007	1.008	1.009
9.9	-	-	1.000	1.002	1.002	1.003	1.004	1.005	1.006	1.008
10.2	-	-	-	1.000	1.001	1.002	1.003	1.004	1.005	1.006
10.5	-	-	-	-	1.000	1.001	1.002	1.003	1.004	1.005
10.8	-	-	-	-	-	1.000	1.001	1.002	1.003	1.003
11.1	-	-	-	-	-	-	1.000	1.001	1.002	1.003
11.4	-	-	-	-	-	-	-	1.000	1.001	1.002
11.7	-	-	-	-	-	-	-	-	1.000	1.001

Table 36: Target at boundary, rework at 20% of nominal

	Elbow Room in std. dev. units				
DNS	***15.***	***18.***	***24.***	***36.***	***60.***
0.0	1.19	1.19	1.19	1.20	1.20
0.3	1.19	1.19	1.19	1.19	1.20
0.6	1.18	1.18	1.19	1.19	1.20
0.9	1.17	1.18	1.18	1.19	1.19
1.2	1.17	1.17	1.18	1.19	1.19
1.5	1.16	1.17	1.18	1.18	1.19
1.8	1.16	1.16	1.17	1.18	1.19
2.1	1.15	1.16	1.17	1.18	1.19
2.4	1.14	1.15	1.16	1.17	1.18
2.7	1.14	1.15	1.16	1.17	1.18
3.0	1.13	1.14	1.15	1.17	1.18
3.3	1.12	1.13	1.15	1.17	1.18
3.6	1.12	1.13	1.14	1.16	1.18
3.9	1.11	1.12	1.14	1.16	1.17
4.2	1.10	1.12	1.14	1.16	1.17
4.5	1.10	1.11	1.13	1.15	1.17
4.8	1.09	1.11	1.13	1.15	1.17
5.1	1.09	1.10	1.12	1.15	1.17
5.4	1.08	1.10	1.12	1.14	1.17
5.7	1.08	1.09	1.12	1.14	1.16
6.0	1.07	1.09	1.11	1.14	1.16
7.5	1.05	1.07	1.09	1.13	1.15
9.0	1.03	1.05	1.08	1.11	1.14
12.	1.01	1.02	1.05	1.09	1.13
15.	1.00	1.01	1.03	1.07	1.11
18.	-	1.00	1.01	1.05	1.10
24.	-	-	1.00	1.02	1.17
36.	-	-	-	1.00	1.03
60.	-	-	-	-	1.00

Table 37: The Effective Cost of Production when the target is at a boundary and nonconforming units are reworked at 10% of nominal

Elbow Room in standard deviation units

DNS	*0.3*	*0.6*	*0.9*	*1.2*	*1.5*	*1.8*	*2.1*	*2.4*	*2.7*	*3.0*
0.0	1.026	1.040	1.050	1.058	1.063	1.068	1.071	1.074	1.077	1.079
0.3	1.000	1.017	1.031	1.041	1.049	1.055	1.060	1.064	1.067	1.070
0.6	-	1.000	1.013	1.025	1.034	1.042	1.048	1.053	1.057	1.061
0.9	-	-	1.000	1.010	1.020	1.029	1.036	1.042	1.046	1.051
1.2	-	-	-	1.000	1.008	1.017	1.024	1.031	1.036	1.040
1.5	-	-	-	-	1.000	1.007	1.014	1.021	1.026	1.031
1.8	-	-	-	-	-	1.000	1.006	1.012	1.018	1.023
2.1	-	-	-	-	-	-	1.000	1.005	1.011	1.015
2.4	-	-	-	-	-	-	-	1.000	1.005	1.009
2.7	-	-	-	-	-	-	-	-	1.000	1.004
3.0	-	-	-	-	-	-	-	-	-	1.000

Elbow Room in standard deviation units

DNS	*3.3*	*3.6*	*3.9*	*4.2*	*4.5*	*4.8*	*5.1*	*5.4*	*5.7*	*6.0*
0.0	1.080	1.082	1.083	1.084	1.085	1.086	1.086	1.087	1.088	1.088
0.3	1.073	1.074	1.076	1.078	1.079	1.080	1.081	1.082	1.083	1.084
0.6	1.064	1.066	1.068	1.070	1.072	1.074	1.075	1.076	1.077	1.078
0.9	1.054	1.057	1.060	1.062	1.064	1.066	1.068	1.069	1.071	1.072
1.2	1.044	1.048	1.051	1.054	1.056	1.058	1.060	1.062	1.064	1.065
1.5	1.035	1.039	1.042	1.045	1.048	1.050	1.053	1.055	1.057	1.058
1.8	1.027	1.031	1.034	1.037	1.040	1.043	1.045	1.047	1.050	1.051
2.1	1.020	1.023	1.027	1.030	1.033	1.036	1.038	1.041	1.043	1.045
2.4	1.013	1.017	1.020	1.023	1.026	1.029	1.032	1.034	1.037	1.039
2.7	1.008	1.012	1.015	1.018	1.021	1.023	1.026	1.028	1.031	1.033
3.0	1.004	1.007	1.010	1.013	1.016	1.018	1.021	1.023	1.025	1.028
3.3	1.000	1.003	1.007	1.009	1.012	1.014	1.016	1.019	1.021	1.023
3.6	-	1.000	1.003	1.006	1.008	1.010	1.012	1.015	1.017	1.019
3.9	-	-	1.000	1.003	1.005	1.007	1.009	1.011	1.013	1.015
4.2	-	-	-	1.000	1.002	1.005	1.007	1.008	1.010	1.012
4.5	-	-	-	-	1.000	1.002	1.004	1.006	1.007	1.009
4.8	-	-	-	-	-	1.000	1.002	1.004	1.005	1.007
5.1	-	-	-	-	-	-	1.000	1.002	1.004	1.005
5.4	-	-	-	-	-	-	-	1.000	1.002	1.003
5.7	-	-	-	-	-	-	-	-	1.000	1.002
6.0	-	-	-	-	-	-	-	-	-	1.000

Table 37: Target at boundary, rework at 10% of nominal

	Elbow Room in standard deviation units									
DNS	***6.3***	***6.6***	***6.9***	***7.2***	***7.5***	***7.8***	***8.1***	***8.4***	***8.7***	***9.0***
0.0	1.089	1.089	1.090	1.090	1.091	1.091	1.091	1.092	1.092	1.092
0.3	1.085	1.085	1.086	1.086	1.087	1.087	1.088	1.088	1.089	1.089
0.6	1.079	1.080	1.081	1.082	1.082	1.083	1.084	1.084	1.085	1.085
0.9	1.073	1.074	1.075	1.076	1.077	1.078	1.079	1.079	1.080	1.081
1.2	1.067	1.068	1.069	1.070	1.071	1.072	1.073	1.074	1.075	1.076
1.5	1.060	1.061	1.063	1.064	1.065	1.067	1.068	1.069	1.070	1.070
1.8	1.053	1.055	1.057	1.058	1.059	1.061	1.062	1.063	1.064	1.065
2.1	1.047	1.049	1.050	1.052	1.054	1.055	1.056	1.058	1.059	1.060
2.4	1.041	1.043	1.045	1.046	1.048	1.050	1.051	1.052	1.054	1.055
2.7	1.035	1.037	1.039	1.041	1.043	1.044	1.046	1.047	1.049	1.050
3.0	1.030	1.032	1.034	1.036	1.038	1.040	1.041	1.043	1.044	1.046
3.3	1.025	1.027	1.029	1.031	1.033	1.035	1.037	1.038	1.040	1.041
3.6	1.021	1.023	1.025	1.027	1.029	1.031	1.032	1.034	1.036	1.037
3.9	1.017	1.019	1.021	1.023	1.025	1.027	1.028	1.030	1.032	1.033
4.2	1.014	1.016	1.017	1.019	1.021	1.023	1.025	1.026	1.028	1.030
4.5	1.011	1.012	1.014	1.016	1.018	1.020	1.021	1.023	1.025	1.026
4.8	1.008	1.010	1.011	1.013	1.015	1.016	1.018	1.020	1.021	1.023
5.1	1.006	1.007	1.009	1.010	1.012	1.014	1.015	1.017	1.018	1.020
5.4	1.004	1.006	1.007	1.008	1.010	1.011	1.013	1.014	1.016	1.017
5.7	1.003	1.004	1.005	1.006	1.008	1.009	1.010	1.012	1.013	1.015
6.0	1.002	1.003	1.004	1.005	1.006	1.007	1.008	1.010	1.011	1.012
6.3	1.000	1.001	1.003	1.003	1.004	1.005	1.006	1.008	1.009	1.010
6.6	-	1.000	1.001	1.002	1.003	1.004	1.005	1.006	1.007	1.008
6.9	-	-	1.000	1.001	1.002	1.003	1.004	1.005	1.006	1.007
7.2	-	-	-	1.000	1.001	1.002	1.003	1.003	1.004	1.005
7.5	-	-	-	-	1.000	1.001	1.002	1.003	1.003	1.004
7.8	-	-	-	-	-	1.000	1.001	1.002	1.002	1.003
8.1	-	-	-	-	-	-	1.000	1.001	1.002	1.002
8.4	-	-	-	-	-	-	-	1.000	1.001	1.002
8.7	-	-	-	-	-	-	-	-	1.000	1.001
9.0	-	-	-	-	-	-	-	-	-	1.000

If you have a negative *DNS* value, use Table 16 to obtain an approximate Effective Cost of Production. To do this divide the *DNS* value by 3 and divide the *Elbow Room* value by 3 (yes 3) to obtain the corresponding values for C_{pk} and C_p to use with Table 16 and look up the Effective Cost.

Table 37: Target at boundary, rework at 10% of nominal

DNS	*Elbow Room in standard deviation units* 9.3	9.6	9.9	10.2	10.5	10.8	11.1	11.4	11.7	12.0
0.0	1.092	1.093	1.093	1.093	1.093	1.093	1.094	1.094	1.094	1.094
0.3	1.089	1.090	1.090	1.090	1.090	1.091	1.091	1.091	1.091	1.092
0.6	1.086	1.086	1.086	1.087	1.087	1.087	1.088	1.088	1.088	1.089
0.9	1.081	1.082	1.082	1.083	1.083	1.084	1.084	1.084	1.085	1.085
1.2	1.076	1.077	1.078	1.078	1.079	1.079	1.080	1.080	1.081	1.081
1.5	1.071	1.072	1.073	1.074	1.074	1.075	1.075	1.076	1.077	1.077
1.8	1.066	1.067	1.068	1.069	1.070	1.070	1.071	1.072	1.072	1.073
2.1	1.061	1.062	1.063	1.064	1.065	1.066	1.067	1.067	1.068	1.069
2.4	1.056	1.057	1.058	1.059	1.060	1.061	1.062	1.063	1.064	1.065
2.7	1.052	1.053	1.054	1.055	1.056	1.057	1.058	1.059	1.060	1.061
3.0	1.047	1.048	1.050	1.051	1.052	1.053	1.054	1.055	1.056	1.057
3.3	1.043	1.044	1.045	1.047	1.048	1.049	1.050	1.051	1.052	1.053
3.6	1.039	1.040	1.042	1.043	1.044	1.045	1.046	1.048	1.049	1.050
3.9	1.035	1.036	1.038	1.039	1.040	1.042	1.043	1.044	1.045	1.046
4.2	1.031	1.033	1.034	1.036	1.037	1.038	1.039	1.041	1.042	1.043
4.5	1.028	1.029	1.031	1.032	1.034	1.035	1.036	1.037	1.039	1.040
4.8	1.025	1.026	1.028	1.029	1.030	1.032	1.033	1.034	1.036	1.037
5.1	1.022	1.023	1.025	1.026	1.027	1.029	1.030	1.031	1.033	1.034
5.4	1.019	1.020	1.022	1.023	1.024	1.026	1.027	1.028	1.030	1.031
5.7	1.016	1.018	1.019	1.020	1.022	1.023	1.024	1.026	1.027	1.028
6.0	1.014	1.015	1.017	1.018	1.019	1.021	1.022	1.023	1.024	1.026
6.3	1.012	1.013	1.014	1.016	1.017	1.018	1.020	1.021	1.022	1.023
6.6	1.010	1.011	1.012	1.013	1.015	1.016	1.017	1.018	1.020	1.021
6.9	1.008	1.009	1.010	1.011	1.013	1.014	1.015	1.016	1.018	1.019
7.2	1.006	1.007	1.008	1.010	1.011	1.012	1.013	1.0.14	1.016	1.017
7.5	1.005	1.006	1.007	1.008	1.009	1.010	1.011	1.012	1.014	1.015
7.8	1.004	1.005	1.006	1.006	1.008	1.009	1.010	1.011	1.012	1.013
8.1	1.003	1.004	1.004	1.005	1.006	1.007	1.008	1.009	1.010	1.011
8.4	1.002	1.003	1.003	1.004	1.005	1.006	1.007	1.008	1.009	1.010
8.7	1.002	1.002	1.002	1.003	1.004	1.005	1.005	1.006	1.007	1.008
9.0	1.001	1.001	1.002	1.002	1.003	1.004	1.004	1.005	1.006	1.007
9.3	1.000	1.001	1.001	1.002	1.002	1.003	1.003	1.004	1.005	1.006
9.6	-	1.000	1.001	1.001	1.002	1.002	1.003	1.003	1.004	1.005
9.9	-	-	1.000	1.001	1.001	1.002	1.002	1.003	1.003	1.004
10.2	-	-	-	1.000	1.001	1.001	1.001	1.002	1.002	1.003
10.5	-	-	-	-	1.000	1.001	1.001	1.001	1.002	1.002
10.8	-	-	-	-	-	1.000	1.001	1.001	1.001	1.002
11.1	-	-	-	-	-	-	1.000	1.001	1.001	1.001
11.4	-	-	-	-	-	-	-	1.000	1.001	1.001
11.7	-	-	-	-	-	-	-	-	1.000	1.001

Table 37: Target at boundary, rework at 10% of nominal

	Elbow Room in std. dev. units				
DNS	***15.***	***18.***	***24.***	***36.***	***60.***
0.0	1.10	1.10	1.10	1.10	1.10
0.3	1.09	1.09	1.10	1.10	1.10
0.6	1.09	1.09	1.09	1.10	1.10
0.9	1.09	1.09	1.09	1.09	1.10
1.2	1.08	1.09	1.09	1.09	1.10
1.5	1.08	1.08	1.09	1.09	1.10
1.8	1.08	1.08	1.09	1.09	1.09
2.1	1.07	1.08	1.08	1.09	1.09
2.4	1.07	1.08	1.08	1.09	1.09
2.7	1.07	1.07	1.08	1.09	1.09
3.0	1.06	1.07	1.08	1.08	1.09
3.3	1.06	1.07	1.07	1.08	1.09
3.6	1.06	1.06	1.07	1.08	1.09
3.9	1.06	1.06	1.07	1.08	1.09
4.2	1.05	1.06	1.07	1.08	1.09
4.5	1.05	1.06	1.07	1.08	1.09
4.8	1.05	1.05	1.06	1.08	1.08
5.1	1.04	1.05	1.06	1.07	1.08
5.4	1.04	1.05	1.06	1.07	1.08
5.7	1.04	1.05	1.06	1.07	1.08
6.0	1.04	1.04	1.06	1.07	1.08
7.5	1.03	1.03	1.05	1.06	1.08
9.0	1.02	1.03	1.04	1.06	1.07
12.	1.00	1.01	1.03	1.04	1.06
15.	1.00	1.00	1.01	1.03	1.06
18.	-	1.00	1.01	1.03	1.05
24.	-	-	1.00	1.01	1.04
36.	-	-	-	1.00	1.02
60.	-	-	-	-	1.00

Tables 38 to 45
Other Tables

Table 38
d_2 Bias Correction Factors for Ranges and Average Ranges

The entries are d_2 values where d_2 = the average number of standard deviation units represented by the range of n data

n	*0*	*1*	2	3	4	5	6	7	8	9
0	—	—	1.128	1.693	2.059	2.326	2.534	2.704	2.847	2.970
10	3.078	3.173	3.258	3.336	3.407	3.472	3.532	3.588	3.640	3.689
20	3.735	3.778	3.819	3.858	3.895	3.931	3.964	3.996	4.027	4.057
30	4.086	4.113	4.139	4.165	4.189	4.213	4.236	4.259	4.280	4.301
40	4.322	4.341	4.361	4.379	4.398	4.415	4.433	4.450	4.466	4.482
50	4.498	4.514	4.529	4.543	4.558	4.572	4.586	4.599	4.613	4.626
60	4.639	4.651	4.663	4.676	4.687	4.699	4.711	4.722	4.733	4.744
70	4.755	4.765	4.776	4.786	4.796	4.806	4.816	4.825	4.835	4.844
80	4.854	4.863	4.872	4.881	4.889	4.898	4.906	4.915	4.923	4.931
90	4.939	4.947	4.955	4.963	4.971	4.978	4.985	4.993	5.001	5.008
100	5.015	5.022	5.029	5.036	5.043	5.050	5.057	5.063	5.070	5.076
110	5.083	5.089	5.096	5.102	5.108	5.114	5.120	5.126	5.132	5.138
120	5.144	5.150	5.156	5.161	5.167	5.173	5.178	5.184	5.189	5.195
130	5.200	5.205	5.211	5.216	5.221	5.226	5.231	5.236	5.241	5.246
140	5.251	5.256	5.261	5.266	5.271	5.275	5.280	5.285	5.289	5.294
150	5.298	5.303	5.308	5.312	5.316	5.321	5.325	5.330	5.334	5.338
160	5.342	5.347	5.351	5.355	5.359	5.363	5.367	5.371	5.375	5.379
170	5.383	5.387	5.391	5.395	5.399	5.403	5.407	5.411	5.414	5.418
180	5.422	5.426	5.429	5.433	5.437	5.440	5.444	5.447	5.451	5.454
190	5.458	5.461	5.465	5.468	5.472	5.475	5.479	5.482	5.485	5.489
200	5.492	5.495	5.499	5.502	5.505	5.508	5.512	5.515	5.518	5.521
210	5.524	5.527	5.531	5.534	5.537	5.540	5.543	5.546	5.549	5.552
220	5.555	5.557	5.561	5.564	5.567	5.570	5.573	5.576	5.578	5.581
230	5.584	5.587	5.590	5.593	5.595	5.598	5.601	5.604	5.606	5.609
240	5.612	5.615	5.617	5.620	5.623	5.626	5.628	5.631	5.633	5.636
250	5.638	5.641	5.644	5.646	5.649	5.651	5.654	5.656	5.659	5.661
260	5.664	5.666	5.669	5.671	5.674	5.676	5.678	5.681	5.683	5.686
270	5.688	5.690	5.693	5.695	5.698	5.700	5.702	5.705	5.707	5.709
280	5.711	5.714	5.716	5.718	5.721	5.723	5.725	5.727	5.729	5.732
290	5.734	5.736	5.738	5.740	5.743	5.745	5.747	5.749	5.751	5.753

Table 38
d_2 Bias Correction Factors for Ranges and Average Ranges

The entries are d_2 values where d_2 = the average number of standard deviation units represented by the range of n data

n	*0*	*1*	*2*	*3*	*4*	*5*	*6*	*7*	*8*	*9*
300	5.756	5.758	5.760	5.762	5.764	5.766	5.768	5.770	5.772	5.774
310	5.776	5.778	5.780	5.783	5.785	5.787	5.789	5.791	5.793	5.795
320	5.797	5.799	5.800	5.802	5.804	5.806	5.808	5.810	5.812	5.814
330	5.816	5.818	5.820	5.822	5.824	5.825	5.827	5.829	5.831	5.833
340	5.835	5.837	5.838	5.840	5.842	5.844	5.846	5.848	5.849	5.851
350	5.853	5.855	5.857	5.858	5.860	5.862	5.864	5.865	5.867	5.869
360	5.871	5.872	5.874	5.876	5.878	5.879	5.881	5.883	5.884	5.886
370	5.887	5.890	5.891	5.893	5.895	5.896	5.898	5.900	5.901	5.903
380	5.904	5.906	5.908	5.909	5.911	5.913	5.914	5.916	5.917	5.919
390	5.921	5.922	5.924	5.925	5.927	5.929	5.930	5.932	5.933	5.935
400	5.936	5.938	5.939	5.941	5.943	5.944	5.946	5.947	5.949	5.950
410	5.952	5.953	5.955	5.956	5.958	5.959	5.961	5.962	5.964	5.965
420	5.967	5.968	5.969	5.971	5.972	5.974	5.975	5.977	5.978	5.980
430	5.981	5.983	5.984	5.985	5.987	5.988	5.990	5.991	5.992	5.994
440	5.995	5.997	5.998	5.999	6.001	6.002	6.004	6.005	6.006	6.008
450	6.009	6.010	6.012	6.013	6.014	6.016	6.017	6.019	6.020	6.021
460	6.023	6.024	6.025	6.026	6.028	6.029	6.030	6.032	6.033	6.034
470	6.036	6.037	6.038	6.040	6.041	6.042	6.043	6.045	6.046	6.047
480	6.049	6.050	6.051	6.052	6.054	6.055	6.056	6.057	6.059	6.060
490	6.061	6.062	6.064	6.065	6.066	6.067	6.069	6.070	6.071	6.072

n	*0*	*10*	*20*	*30*	*40*	*50*	*60*	*70*	*80*	*90*
500	6.073	6.085	6.097	6.109	6.120	6.131	6.142	6.153	6.163	6.173
600	6.183	6.193	6.203	6.213	6.222	6.231	6.240	6.249	6.258	6.267
700	6.275	6.283	6.292	6.300	6.308	6.316	6.324	6.331	6.339	6.346
800	6.354	6.361	6.368	6.375	6.382	6.389	6.396	6.402	6.409	6.416
900	6.422	6.429	6.435	6.441	6.447	6.453	6.459	6.465	6.471	6.477
1000	6.483									

Table 38
d_2 Bias Correction Factors for Ranges and Average Ranges

The values for d_2 are defined by the following expression:

$$d_2 = \int_0^{\infty} w \, \frac{n\,(n-1)}{[\sqrt{2\pi}\,]^n} \int_{-\infty}^{\infty} \left[\int_x^{x+w} e^{-(u^2/2)} \, du \right]^{n-2} e^{-((x+w)^2/2)} \; e^{-(x^2/2)} \; dx \; dw$$

(Other bias correction factors are given in Table 45.)

Table 39: Charts for Individual Values and Moving Ranges

When a *time series* has a logical subgroup size of $n = 1$ we can plot the individual values and use a two-point moving range to measure the dispersion. The Average Moving Range or the Median Moving Range may then be used to compute the Natural Process Limits for the individual values and the Upper Range Limit for the moving ranges according to the formulas:

$$UNPL_X = \bar{X} + E_2\,\bar{R} \quad \text{or} \quad \bar{X} + E_5\,\tilde{R}$$
$$CL_X = \bar{X}$$
$$LNPL_X = \bar{X} - E_2\,\bar{R} \quad \text{or} \quad \bar{X} - E_5\,\tilde{R}$$

$$URL_R = D_4\,\bar{R} \quad \text{or} \quad D_6\,\tilde{R}$$
$$CL_R = \bar{R} \quad \text{or} \quad \tilde{R}$$

Table 39: Charts for Individual Values and Moving Ranges

For two-period moving ranges these constants are:

$$E_2 = \frac{3}{d_2} = 2.660 \qquad E_5 = \frac{3}{d_4} = 3.145$$

$$D_4 = \left[1 + \frac{3\,d_3}{d_2} \right] = 3.268 \qquad D_6 = \frac{d_2 + 3\,d_3}{d_4} = 3.865$$

where d_3 and d_4 are bias correction factors given in Table 45.

Moving ranges which fall above the upper range limit may be taken as indications of a potential break in the time series—some sudden shift which is so large that it is unlikely to have occurred by chance.

There is no advantage to using any other measure of dispersion besides a two-point moving range. Moving ranges based on any $n > 2$ will be harder to compute and less efficient than the two-point moving ranges. Mean square successive differences will be harder to compute and more prone to being inflated by extreme values while being no more efficient than the two-point moving range. Global measures of dispersion beg the question of statistical control by making a strong assumption that the data are completely homogeneous, and so they can never be properly used to compute limits for individual values.

If the data do not come from a time series, i.e., if the individual values do not possess a definite and known time order, then the use of a moving range to measure the dispersion of the data is essentially arbitrary because the order of the data is arbitrary.

Moreover, if the individual values are *known* to come from different cause systems, as would be the case in a sequence of experimental runs, then the moving ranges may not be used to construct control limits even when the data constitute a time series.

Table 40: Average and Range Charts
Factors for Using the Average Range, $\bar{R}$

Given k subgroups each with n observations
with Grand Average, $\bar{\bar{X}}$ and Average Range, $\bar{R}$
use the tabled constants with the formulas to obtain

- limits for Subgroup Averages,
- limits for Subgroup Ranges, or
- Natural Process Limits for X.

n	A_2	D_3	D_4	E_2
2	1.880	--	3.268	2.660
3	1.023	--	2.574	1.772
4	0.729	--	2.282	1.457
5	0.577	--	2.114	1.290
6	0.483	--	2.004	1.184
7	0.419	0.076	1.924	1.109
8	0.373	0.136	1.864	1.054
9	0.337	0.184	1.816	1.010
10	0.308	0.223	1.777	0.975
11	0.285	0.256	1.744	0.945
12	0.266	0.283	1.717	0.921
13	0.249	0.307	1.693	0.899
14	0.235	0.328	1.672	0.881
15	0.223	0.347	1.653	0.864

Table 40: Average and Range Charts Factors for Using the Average Range, $\bar{R}$

Limits for Subgroup Averages will be:

$$UAL_{\bar{X}} = \bar{\bar{X}} + A_2 \bar{R}$$
$$CL_{\bar{X}} = \bar{\bar{X}}$$
$$LAL_{\bar{X}} = \bar{\bar{X}} - A_2 \bar{R}$$

Limits for Subgroup Ranges will be:

$$URL_R = D_4 \bar{R}$$
$$CL_R = \bar{R}$$
$$LRL_R = D_3 \bar{R}$$

While Natural Process Limits for individual values may be obtained from:

$$UNPL_X = \bar{\bar{X}} + E_2 \bar{R}$$
$$CL_X = \bar{\bar{X}}$$
$$LNPL_X = \bar{\bar{X}} - E_2 \bar{R}$$

where

$$A_2 = \frac{3}{d_2 \sqrt{n}}$$

$$D_3 = \left[1 - \frac{3\,d_3}{d_2} \right] \qquad D_4 = \left[1 + \frac{3\,d_3}{d_2} \right]$$

$$E_2 = \frac{3}{d_2}$$

Table 41: Average and Range Charts
Factors for Using the Median Range, $\tilde{R}$

Given k subgroups each with n observations
with Grand Average, $\bar{\bar{X}}$ and Median Range, $\tilde{R}$
use the tabled constants with the formulas below to obtain

- limits for Subgroup Averages,
- limits for Subgroup Ranges, or
- Natural Process Limits for X.

n	A_4	D_5	D_6	E_5
2	2.224	--	3.865	3.145
3	1.091	--	2.745	1.889
4	0.758	--	2.375	1.517
5	0.594	--	2.179	1.329
6	0.495	--	2.055	1.214
7	0.429	0.078	1.967	1.134
8	0.380	0.139	1.901	1.075
9	0.343	0.187	1.850	1.029
10	0.314	0.227	1.809	0.992
11	0.290	0.260	1.773	0.961
12	0.270	0.288	1.744	0.935
13	0.253	0.312	1.719	0.913
14	0.239	0.333	1.697	0.894
15	0.226	0.352	1.678	0.877

Table 41: Average and Range Charts Factors for Using the Median Range, $\tilde{R}$

Limits for Subgroup Averages will be:

$$UAL_{\bar{X}} = \bar{\bar{X}} + A_4\,\tilde{R}$$
$$CL_{\bar{X}} = \bar{\bar{X}}$$
$$LAL_{\bar{X}} = \bar{\bar{X}} - A_4\,\tilde{R}$$

Limits for Subgroup Ranges will be:

$$URL_R = D_6\,\tilde{R}$$
$$CL_R = \tilde{R}$$
$$LRL_R = D_5\,\tilde{R}$$

While Natural Process Limits for individual values may be obtained from:

$$UNPL_X = \bar{\bar{X}} + E_5\,\tilde{R}$$
$$CL_X = \bar{\bar{X}}$$
$$LNPL_X = \bar{\bar{X}} - E_5\,\tilde{R}$$

where

$$A_4 = \frac{3}{d_4\sqrt{n}}$$

$$D_5 = \frac{d_2 - 3\,d_3}{d_4} \qquad D_6 = \frac{d_2 + 3\,d_3}{d_4}$$

$$E_5 = \frac{3}{d_4}$$

Table 42: Average and Standard Deviation Charts Factors for Using the Average Standard Deviation, $\bar{s}$

Given k subgroups each with n observations with Grand Average, $\bar{\bar{X}}$, and Average Standard Deviation, $\bar{s}$ use the tabled constants with the formulas below to obtain

- limits for Subgroup Averages,
- limits for Subgroup Standard Deviations, or
- Natural Process Limits for individual values.

n	A_3	B_3	B_4	E_3
2	2.659	--	3.267	3.760
3	1.954	--	2.568	3.385
4	1.628	--	2.266	3.256
5	1.427	--	2.089	3.191
6	1.287	0.030	1.970	3.153
7	1.182	0.118	1.882	3.127
8	1.099	0.185	1.815	3.109
9	1.032	0.239	1.761	3.095
10	0.975	0.284	1.716	3.084
11	0.927	0.322	1.678	3.076
12	0.886	0.354	1.646	3.069
13	0.850	0.382	1.619	3.063
14	0.817	0.407	1.593	3.058
15	0.789	0.428	1.572	3.054

Table 42: Average and Standard Deviation Charts Factors for Using the Average Standard Deviation, $\bar{s}$

Limits for Subgroup Averages will be:

$$UAL_{\bar{X}} = \bar{\bar{X}} + A_3\,\bar{s}$$
$$CL_{\bar{X}} = \bar{\bar{X}}$$
$$LAL_{\bar{X}} = \bar{\bar{X}} - A_3\,\bar{s}$$

Limits for Subgroup Standard Deviations will be:

$$USDL_S = B_4\,\bar{s}$$
$$CL_S = \bar{s}$$
$$LSDL_S = B_3\,\bar{s}$$

While Natural Process Limits for individual values may be obtained from:

$$UNPL_X = \bar{\bar{X}} + E_3\,\bar{s}$$
$$CL_X = \bar{\bar{X}}$$
$$LNPL_X = \bar{\bar{X}} - E_3\,\bar{s}$$

where

$$A_3 = \frac{3}{c_4\sqrt{n}}$$

$$B_3 = 1 - \frac{3}{c_4}\sqrt{1-(c_4)^2} \qquad B_4 = 1 + \frac{3}{c_4}\sqrt{1-(c_4)^2}$$

$$E_3 = \frac{3}{c_4}$$

Table 43: Average and Standard Deviation Charts Factors for Using the Median Standard Deviation, $\tilde{s}$

Given k subgroups each with n observations
with Grand Average, $\bar{\bar{X}}$, and Median Standard Deviation, $\tilde{s}$
use the tabled constants with the formulas below to obtain

- limits for Subgroup Averages,
- limits for Subgroup Standard Deviations, or
- Natural Process Limits for individual values.

n	A_{10}	B_9	B_{10}	E_6
2	3.143	--	3.864	4.444
3	2.082	--	2.733	3.606
4	1.689	--	2.351	3.378
5	1.465	--	2.145	3.275
6	1.313	0.031	2.008	3.215
7	1.201	0.120	1.913	3.178
8	1.114	0.188	1.839	3.151
9	1.044	0.242	1.782	3.132
10	0.985	0.287	1.735	3.115
11	0.936	0.324	1.695	3.106
12	0.893	0.357	1.661	3.093
13	0.856	0.384	1.631	3.086
14	0.823	0.409	1.604	3.080
15	0.794	0.431	1.582	3.074

Table 43: Average and Standard Deviation Charts Factors for Using the Median Standard Deviation, $\tilde{s}$

Limits for Subgroup Averages will be:

$$UAL_{\bar{X}} = \bar{\bar{X}} + A_{10}\,\tilde{s}$$

$$CL_{\bar{X}} = \bar{\bar{X}}$$

$$LAL_{\bar{X}} = \bar{\bar{X}} - A_{10}\,\tilde{s}$$

Limits for Subgroup Standard Deviations will be:

$$USDL_S = B_{10}\,\tilde{s}$$

$$CL_S = \tilde{s}$$

$$LSDL_S = B_9\,\tilde{s}$$

While Natural Process Limits for individual values may be obtained from:

$$UNPL_X = \bar{\bar{X}} + E_6\,\tilde{s}$$

$$CL_X = \bar{\bar{X}}$$

$$LNPL_X = \bar{\bar{X}} - E_6\,\tilde{s}$$

where

$$A_{10} = \frac{3}{c_6\sqrt{n}}$$

$$B_9 = \frac{c_4 - 3\sqrt{1-(c_4)^2}}{c_6} \qquad B_{10} = \frac{c_4 + 3\sqrt{1-(c_4)^2}}{c_6}$$

$$E_6 = \frac{3}{c_6}$$

Table 44: Areas Under the Standard Normal Curve

z	0.09	0.08	0.07	0.06	0.05	0.04	0.03	0.02	0.01	0.00
–3.4	.0002	.0003	.0003	.0003	.0003	.0003	.0003	.0003	.0003	.0003
–3.3	.0003	.0004	.0004	.0004	.0004	.0004	.0004	.0005	.0005	.0005
–3.2	.0005	.0005	.0005	.0006	.0006	.0006	.0006	.0006	.0007	.0007
–3.1	.0007	.0007	.0008	.0008	.0008	.0008	.0009	.0009	.0009	.0010
–3.0	.0010	.0010	.0011	.0011	.0011	.0012	.0012	.0013	.0013	.0013
–2.9	.0014	.0014	.0015	.0015	.0016	.0016	.0017	.0018	.0018	.0019
–2.8	.0019	.0020	.0021	.0021	.0022	.0023	.0023	.0024	.0025	.0026
–2.7	.0026	.0027	.0028	.0029	.0030	.0031	.0032	.0033	.0034	.0035
–2.6	.0036	.0037	.0038	.0039	.0040	.0041	.0043	.0044	.0045	.0047
–2.5	.0048	.0049	.0051	.0052	.0054	.0055	.0057	.0059	.0060	.0062
–2.4	.0064	.0066	.0068	.0069	.0071	.0073	.0075	.0078	.0080	.0082
–2.3	.0084	.0087	.0089	.0091	.0094	.0096	.0099	.0102	.0104	.0107
–2.2	.0110	.0113	.0116	.0119	.0122	.0125	.0129	.0132	.0136	.0139
–2.1	.0143	.0146	.0150	.0154	.0158	.0162	.0166	.0170	.0174	.0179
–2.0	.0183	.0188	.0192	.0197	.0202	.0207	.0212	.0217	.0222	.0228
–1.9	.0233	.0239	.0244	.0250	.0256	.0262	.0268	.0274	.0281	.0287
–1.8	.0294	.0301	.0307	.0314	.0322	.0329	.0336	.0344	.0352	.0359
–1.7	.0367	.0375	.0384	.0392	.0401	.0409	.0418	.0427	.0436	.0446
–1.6	.0455	.0465	.0475	.0485	.0495	.0505	.0516	.0526	.0537	.0548
–1.5	.0559	.0571	.0582	.0594	.0606	.0618	.0630	.0643	.0655	.0668
–1.4	.0681	.0694	.0708	.0722	.0735	.0749	.0764	.0778	.0793	.0808
–1.3	.0823	.0838	.0853	.0869	.0885	.0901	.0918	.0934	.0951	.0968
–1.2	.0985	.1003	.1020	.1038	.1056	.1075	.1093	.1112	.1131	.1151
–1.1	.1170	.1190	.1210	.1230	.1251	.1271	.1292	.1314	.1335	.1357
–1.0	.1379	.1401	.1423	.1446	.1469	.1492	.1515	.1539	.1562	.1587
–0.9	.1611	.1635	.1660	.1685	.1711	.1736	.1762	.1788	.1814	.1841
–0.8	.1867	.1894	.1922	.1949	.1977	.2005	.2033	.2061	.2090	.2119
–0.7	.2148	.2177	.2206	.2236	.2266	.2296	.2327	.2358	.2389	.2420
–0.6	.2451	.2483	.2514	.2546	.2578	.2611	.2643	.2676	.2709	.2743
–0.5	.2776	.2810	.2843	.2877	.2912	.2946	.2981	.3015	.3050	.3085
–0.4	.3121	.3156	.3192	.3228	.3264	.3300	.3336	.3372	.3409	.3446
–0.3	.3483	.3520	.3557	.3594	.3632	.3669	.3707	.3745	.3783	.3821
–0.2	.3859	.3897	.3936	.3974	.4013	.4052	.4090	.4129	.4168	.4207
–0.1	.4247	.4286	.4325	.4364	.4404	.4443	.4483	.4522	.4562	.4602
–0.0	.4641	.4681	.4721	.4761	.4801	.4840	.4880	.4920	.4960	.5000

Table 44: Areas Under the Standard Normal Curve

z	**0.00**	**0.01**	**0.02**	**0.03**	**0.04**	**0.05**	**0.06**	**0.07**	**0.08**	**0.09**
0.0	.5000	.5040	.5080	.5120	.5160	.5199	.5239	.5279	.5319	.5359
0.1	.5398	.5438	.5478	.5517	.5557	.5596	.5636	.5675	.5714	.5753
0.2	.5793	.5832	.5871	.5910	.5948	.5987	.6026	.6064	.6103	.6141
0.3	.6179	.6217	.6255	.6293	.6331	.6368	.6406	.6443	.6480	.6517
0.4	.6554	.6591	.6628	.6664	.6700	.6736	.6772	.6808	.6844	.6879
0.5	.6915	.6950	.6985	.7019	.7054	.7088	.7123	.7157	.7190	.7224
0.6	.7257	.7291	.7324	.7357	.7389	.7422	.7454	.7486	.7517	.7549
0.7	.7580	.7611	.7642	.7673	.7704	.7734	.7764	.7794	.7823	.7852
0.8	.7881	.7910	.7939	.7967	.7995	.8023	.8051	.8078	.8106	.8133
0.9	.8159	.8186	.8212	.8238	.8264	.8289	.8315	.8340	.8365	.8389
1.0	.8413	.8438	.8461	.8485	.8508	.8531	.8554	.8577	.8599	.8621
1.1	.8643	.8665	.8686	.8708	.8729	.8749	.8770	.8790	.8810	.8830
1.2	.8849	.8869	.8888	.8907	.8925	.8944	.8962	.8980	.8997	.9015
1.3	.9032	.9049	.9066	.9082	.9099	.9115	.9131	.9147	.9162	.9177
1.4	.9192	.9207	.6222	.9236	.9251	.9265	.9278	.9292	.9306	.9319
1.5	.9332	.9345	.9357	.9370	.9382	.9394	.9406	.9418	.9429	.9441
1.6	.9452	.9463	.9474	.9484	.9495	.9505	.9515	.9525	.9535	.9545
1.7	.9554	.9564	.9573	.9582	.9591	.9599	.9608	.9616	.9625	.9633
1.8	.9641	.9649	.9656	.9664	.9671	.9678	.9686	.9693	.9699	.9706
1.9	.9713	.9719	.9726	.9732	.9738	.9744	.9750	.9756	.9761	.9767
2.0	.9772	.9778	.9783	.9788	.9793	.9798	.9803	.9808	.9812	.9817
2.1	.9821	.9826	.9830	.9834	.9838	.9842	.9846	.9850	.9854	.9857
2.2	.9861	.9864	.9868	.9871	.9875	.9878	.9881	.9884	.9887	.9890
2.3	.9893	.9896	.9898	.9901	.9904	.9906	.9909	.9911	.9913	.9916
2.4	.9918	.9920	.9922	.9925	.9927	.9929	.9931	.9932	.9934	.9936
2.5	.9938	.9940	.9941	.9943	.9945	.9946	.9948	.9949	.9951	.9952
2.6	.9953	.9955	.9956	.9957	.9959	.9960	.9961	.9962	.9963	.9964
2.7	.9965	.9966	.9967	.9968	.9969	.9970	.9971	.9972	.9973	.9974
2.8	.9974	.9975	.9976	.9977	.9977	.9978	.9979	.9979	.9980	.9981
2.9	.9981	.9982	.9982	.9983	.9984	.9984	.9985	.9985	.9986	.9986
3.0	.9987	.9987	.9987	.9988	.9988	.9989	.9989	.9989	.9990	.9990
3.1	.9990	.9991	.9991	.9991	.9992	.9992	.9992	.9992	.9993	.9993
3.2	.9993	.9993	.9994	.9994	.9994	.9994	.9994	.9995	.9995	.9995
3.3	.9995	.9995	.9995	.9996	.9996	.9996	.9996	.9996	.9996	.9997
3.4	.9997	.9997	.9997	.9997	.9997	.9997	.9997	.9997	.9997	.9998

Table 45: Additional Bias Correction Factors

n	c_4	c_6	d_3	d_4	n	c_4	c_6	d_3	d_4
2	.7979	.675	0.8525	0.954	**21**	.9876	.983	0.7272	3.730
3	.8862	.832	0.8884	1.588	**22**	.9882	.984	0.7199	3.771
4	.9213	.888	0.8798	1.978	**23**	.9887	.985	0.7159	3.811
5	.9400	.916	0.8641	2.257	**24**	.9892	.985	0.7121	3.847
6	.9515	.933	0.8480	2.472	**25**	.9896	.986	0.7084	3.883
7	.9594	.944	0.8332	2.645	**30**	.9915	.988	0.6927	4.037
8	.9650	.952	0.8198	2.791	**35**	.9927	.990	0.6799	4.166
9	.9693	.958	0.8078	2.915	**40**	.9936	.991	0.6692	4.274
10	.9727	.963	0.7971	3.024	**45**	.9943	.992	0.6601	4.372
11	.9754	.966	0.7873	3.121	**50**	.9949	.993	0.6521	4.450
12	.9776	.970	0.7785	3.207	**55**	.9954	.994	0.6452	4.521
13	.9794	.972	0.7704	3.285	**60**	.9957	.994	0.6389	4.591
14	.9810	.974	0.7630	3.356	**65**	.9961	.995	0.6333	4.649
15	.9823	.976	0.7562	3.422	**70**	.9964	.995	0.6283	4.707
16	.9835	.978	0.7499	3.482	**75**	.9966	.996	0.6236	4.757
17	.9845	.979	0.7441	3.538	**80**	.9968	.996	0.6194	4.806
18	.9854	.980	0.7386	3.591	**85**	.9970	.996	0.6154	4.849
19	.9862	.981	0.7335	3.640	**90**	.9972	.996	0.6118	4.892
20	.9869	.982	0.7287	3.686	**100**	.9975	.997	0.6052	4.968

$$c_4 = \sqrt{\frac{2}{n-1}}\,\frac{\Gamma(\frac{n}{2})}{\Gamma(\frac{n-1}{2})}$$

c_6 is defined by the equation: $$0.5\,\Gamma(\tfrac{n-1}{2}) = \int_0^{c_6^2/2(n-1)} t^{(n-3)/2}\, e^{-t}\, dt$$

$$d_3^2 = \int_0^\infty [w - d_2]^2 f(w)\, dw \quad \text{and} \quad 0.5000 = \int_0^{d_4} f(w)\, dw \quad \text{where}$$

$$f(w) = \frac{n\,(n-1)}{[\sqrt{2\pi}\,]^n} \int_{-\infty}^{\infty} \left[\int_x^{x+w} e^{-(u^2/2)}\, du\right]^{n-2} e^{-((x+w)^2/2)}\; e^{-(x^2/2)}\, dx$$

Worksheets

Worksheet for Two-Sided Specifications

page 1 of 4

Part One: Characterizing Past Performance

1. Identify a critical product characteristic:

Product Identification: ______________________________

Critical Characteristic Evaluated: ______________________________

Upper Specification Limit = *USL* = ____________

Target Value = *target* = ____________

Lower Specification Limit = *LSL* = ____________

2. Compute a *scale factor* using a global measure of dispersion:

number of values used = *N* = ______

scale factor: s = ____________ or R/d_2 = ____________

3. Characterize the *Elbow Room*:

specified tolerance = *USL* – *LSL* = ____________

Divide by your *scale factor* to obtain the

Elbow Room = ———————— = ____________

Divide by 6.0 to obtain your

Performance Ratio, P_p = [ ]

4. Characterize the *Distance to Nearer Specification*:

average value for the *N* data = ____________

USL – average = *DUS* = ____________

average – LSL = *DLS* = ____________

$$DNS = \frac{\text{smaller of } DUS \text{ and } DLS}{\textit{scale factor}} = \text{————} = \text{\_\_\_\_\_\_\_\_}$$

Divide the *DNS* value by 3.0 to obtain your

Centered Performance Ratio, P_{pk} = [ ]

Worksheet for Two-Sided Specifications

page 2 of 4

5. Find the Effective Cost of Production: Fill in the blanks below:

Average = ________

scrap? *rework?* *LSL* = _______ *target* = _________ *USL* = _______ *scrap?* *rework?*

What happens to product that exceeds the upper specification?

What happens to product that falls below the lower specification?

What is the cost of rework as a proportion of the cost of scrap?

100% 80% 67% 50% 33% 20% 10% 1% 0.1%

Is the average on the rework side or the scrap side of the target?

Which Table fits your situation? Table ______

From this table, with P_p = , and with P_{pk} =

read off your ***Effective Cost of Production:*** [ ]

The amount by which this value exceeds 1.00 represents the excess costs associated with your past production.

Worksheet for Two-Sided Specifications

page 3 of 4

Part Two: Potential Payoffs for Improvements

6. Find your *Centered Cost of Production*:

Using the table identified in Step 5, and with $P_{pk} = P_p$ find the Effective Cost of Production for a centered process:

Centered Cost of Production = ____________

Subtract the Centered Cost of Production from the Effective Cost of Production to find the

Potential Benefit of Operating On-Target = [ ]

(Multiply by nominal cost of production to get benefit in dollars.)

7. Place your past performance data on a process behavior chart:

☐ If your process appears to be predictable skip Steps 7, 8, and 9. Unless or until your process is changed in some major way, the cost of production in Step 5 is what you should expect from your process in the future. If you improve your process aim you can expect to get the cost of production found in Step 6.

☐ If your process appears to be unpredictable, then you will need to compute a *Potential Cost of Production*. As a first step toward this value use the Average Range, or the Average Moving Range, from your process behavior chart and the appropriate bias correction factor from Table 38 to compute a *predictable scale factor:*

subgroup size, n = ______ (use $n = 2$ for XmR chart)

Average Range = $\bar{R}$ = __________

predictable scale factor = $\frac{\bar{R}}{d_2}$ = __________ = __________

Worksheet for Two-Sided Specifications

page 4 of 4

8. Find your hypothetical process capability:

Divide your specified tolerance in measurement units

by your *predictable scale factor* = ____________ = ____________

Divide this value by 6.0 to obtain your

Hypothetical Capability Ratio, C_p = ____________

9. Find the Potential Cost of Production:

Assume that C_{pk} is equal to C_p

and use the table identified in Step 5

to find the ***Potential Cost of Production***: [ ]

Subtract the Potential Cost of Production

from the Effective Cost of Production found in Step 5

to obtain the minimum potential benefit

to be had by operating predictably and on-target:

Minimum Potential Benefit from Predictable and On-target [ ]

Multiply by the nominal cost of production to turn this into dollars.

Worksheet for One-Sided Specification Without Target

Part One: Characterizing Past Performance

1. Identify a critical product characteristic:

Product Identification: ______________________________

Critical Characteristic Evaluated: ______________________

Specification Limit = SL = __________

2. Compute a *scale factor* using a global measure of dispersion:

Number of values used = N = _____

scale factor: s = __________ or R/d_2 = __________

3. Characterize the Distance to (Nearer) Specification:

average value for the N data = __________

$USL - average$ or $average - LSL$ = __________

$$DNS = \frac{average\ \ to\ spec}{scale\ factor} = \text{\_\_\_\_\_\_\_\_} = \text{\_\_\_\_\_\_\_\_}$$

Divide the DNS value by 3.0 to obtain your

Centered Performance Ratio, P_{pk} = [ ]

4. Find the Effective Cost of Production:

Is nonconforming product scrapped or reworked?

If it is reworked, then what is the cost of rework as a proportion of the nominal cost of production? 100% 80% 67% 50% 33% 20% 10%

Use the appropriate column of Table 21, let P_{pk} determine the row, and read off your ***Effective Cost of Production:*** [ ]

Multiply by the nominal cost of production (at the average) to find the **actual cost of production** for past production [ ]

Worksheet for One-Sided Specification Without Target

page 2 of 5

Part Two: Potential Payoff for Improving Process Aim

5. Find the aim points that correspond to the C_{pk} values of Table 21:

$$aim\ point = specification \pm [\ C_{pk} \times scale\ factor \times 3.0\ ]$$

C_{pk}	Aim Point	Nominal Cost	Effective Cost	Actual Cost
0.0	______	______	______	______
0.1	______	______	______	______
0.2	______	______	______	______
0.3	______	______	______	______
0.4	______	______	______	______
0.5	______	______	______	______
0.6	______	______	______	______
0.7	______	______	______	______
0.8	______	______	______	______
0.9	______	______	______	______
1.0	______	______	______	______
1.1	______	______	______	______
1.2	______	______	______	______
1.3	______	______	______	______
1.4	______	______	______	______
1.5	______	______	______	______
1.6	______	______	______	______
1.7	______	______	______	______

Worksheet for One-Sided Specification Without Target

page 3 of 5

6. **For each of the aim points find the nominal cost of production**
Use your process knowledge to obtain reasonable and appropriate estimates of the nominal cost of production at each one of the different aim points computed in Step 5 above.

7. **Use Table 21 to obtain the Effective Costs of Production**
Use the appropriate column of Table 21 to fill in the Effective Costs of Production that correspond to the C_{pk} values shown in Step 5.

8. **Multiply the nominal costs by the Effective Costs to find the actual costs of production**
Identify the aim point that gives the minimum actual cost.

Optimum aim point = ______________

Effective Cost of Production at optimum aim = ______________

Optimum actual cost of production = ______________

Compare this actual cost with the value found in Step 4 to determine the potential payback from improving the process aim.

Worksheet for One-Sided Specification Without Target

page 4 of 5

Part Three: Potential Payoff for Operating Predictably

9. Place your past performance data on a process behavior chart:

☐ If your process appears to be predictable, skip the remainder of this worksheet. Unless or until your process is changed in some major way, the cost of production in Step 4 is what you should expect from your process in the future. If you improve your process aim you can expect to get the cost of production found in Step 8.

☐ If your process appears to be unpredictable, then you will need to compute a *Potential Cost of Production*. As a first step toward this value use the Average Range, or the Average Moving Range, from your process behavior chart and the appropriate bias correction factor from Table 38 to compute a *predictable scale factor:*

subgroup size, n = ______ (use $n = 2$ for XmR chart)

Average Range = $\bar{R}$ = ____________

$$\textit{predictable scale factor} = \frac{\bar{R}}{d_2} = \_\_\_\_\_\_\_\_\_\_\_\_ = \_\_\_\_\_\_\_\_\_\_\_\_$$

10. Evaluate the benefits of operating predictably at current average:

Use the predictable scale factor above to compute a hypothetical centered capability ratio:

$$\textit{Hypothetical } C_{pk} = \frac{|\,\textit{Average} - \textit{Specification}\,|}{3 \times \textit{predictable scale factor}} = \_\_\_\_\_\_\_\_ = \_\_\_\_\_\_\_\_\_\_$$

Use this value with the appropriate column of Table 21 to find a ***Potential Cost of Production*** = ______________

Compare with the Effective Cost of Production found in Step 4 to determine the benefits of operating this process predictably at the current average.

Worksheet for One-Sided Specification Without Target

11. Evaluate the benefits of operating predictably at optimum aim:

Recompute the aim points using the *predictable scale factor*:

aim point = *specification* ± [C_{pk} X *scale factor* X 3.0]

C_{pk}	Aim Point	Nominal Cost	Effective Cost	Actual Cost
0.0	______	______	______	______
0.1	______	______	______	______
0.2	______	______	______	______
0.3	______	______	______	______
0.4	______	______	______	______
0.5	______	______	______	______
0.6	______	______	______	______
0.7	______	______	______	______
0.8	______	______	______	______
0.9	______	______	______	______
1.0	______	______	______	______
1.1	______	______	______	______
1.2	______	______	______	______
1.3	______	______	______	______
1.4	______	______	______	______
1.5	______	______	______	______
1.6	______	______	______	______
1.7	______	______	______	______

- Find the nominal cost of production at each aim point.
- Find the Effective Costs of Production at each aim point.
- Multiply nominal costs by Effective Costs and find the minimum actual cost.
- Compare this minimum with the actual cost in Step 4 to determine the benefits of operating predictably at optimum aim.

Worksheet for a One-Sided Specification With Target

page 1 of 6

Part One: Characterizing Past Performance

1. Identify a critical product characteristic:

product identification: ______________________

critical characteristic evaluated: ______________________

specification limit = *SL* = ____________

target value = *target* = ____________

2. Compute a *scale factor* using a global measure of dispersion:

number of values used = N = ______

scale factor: s = ____________ or R/d_2 = ____________

3. Characterize the *Tolerance* in standard deviation units:

$$Tolerance = \frac{|\ target - SL\ |}{scale\ factor} = \underline{\qquad\qquad} = \boxed{\qquad\qquad}$$

4. Characterize the Distance to (Nearer) Specification:

average value for the N data = ____________

$USL - average$ or $average - LSL$ = ____________

$$DNS = \frac{average\ \ to\ spec}{scale\ factor} = \underline{\qquad\qquad} = \underline{\qquad\qquad}$$

Divide the *DNS* value by 3.0 to obtain your

Centered Performance Ratio, P_{pk} = $\boxed{\qquad\qquad}$

Worksheet for a One-Sided Specification With Target

page 2 of 6

5. Find the Effective Cost of Production:

Is nonconforming product scrapped or reworked?

If it is reworked, then what is the cost of rework as a proportion of the nominal cost of production? 100% 80% 67% 50% 33% 20% 10%

Find the appropriate table from Tables 22 through 29: Table _____

Use the *Tolerance* value and the P_{pk} value to find your **Effective Cost of Production**: [ ]

Multiply by the Nominal Cost of Production (at the average) to find the **actual cost of production** for past production [ ]

Part Two: Potential Payoff for Improving Process Aim

6. Find aim points that correspond to the C_{pk} values of the table above:

Fill these values in on the following page.

$$\textit{aim point} = \textit{specification} \pm [\, C_{pk} \times \textit{Scale Factor} \times 3.0 \,]$$

7. For each of the aim points find the nominal cost of production
Use your process knowledge to obtain reasonable and appropriate estimates of the nominal cost of production at each of the different aim points computed in Step 6 above.

8. Use the table above to obtain the Effective Costs of Production
Use the appropriate column of the table identified in Step 5 to fill in the Effective Costs of Production that correspond to the C_{pk} values shown on the next page.

Worksheet for a One-Sided Specification With Target

page 3 of 6

C_{pk}	Aim Point	Nominal Cost	Effective Cost	Actual Cost
0.0	______	______	______	______
0.1	______	______	______	______
0.2	______	______	______	______
0.3	______	______	______	______
0.4	______	______	______	______
0.5	______	______	______	______
0.6	______	______	______	______
0.7	______	______	______	______
0.8	______	______	______	______
0.9	______	______	______	______
1.0	______	______	______	______
1.1	______	______	______	______
1.2	______	______	______	______
1.3	______	______	______	______
1.4	______	______	______	______
1.5	______	______	______	______
1.6	______	______	______	______
1.7	______	______	______	______
1.8	______	______	______	______
1.9	______	______	______	______
2.0	______	______	______	______

Worksheet for a One-Sided Specification With Target

9. **On page 3 multiply the nominal costs by the Effective Costs to find the actual costs of production**
Identify the aim point that gives the minimum actual cost.

Optimum aim point = ____________

Optimum actual cost of production = ____________

To determine the potential payback from improving the process aim compare this actual cost with the value found in Step 5.

Part Three: Potential Payoff for Operating Predictably

10. **Place your past performance data on a process behavior chart:**

☐ If your process appears to be predictable, skip the remainder of this worksheet. Unless or until your process is changed in some major way, the cost of production in Step 5 is what you should expect from your process in the future. If you improve your process aim you can expect to get the cost of production found in Step 9.

☐ If your process appears to be unpredictable, then you will need to compute a *Potential Cost of Production*. As a first step toward this value use the Average Range, or the Average Moving Range, from your process behavior chart and the appropriate bias correction factor from Table 38 to compute a *predictable scale factor:*

subgroup size, n = ______ (use $n = 2$ for XmR chart)

Average Range = $\bar{R}$ = ____________

predictable scale factor = $\frac{\bar{R}}{d_2}$ = ____________ = ____________

Worksheet for a One-Sided Specification With Target

page 5 of 6

11. Evaluate the benefits of operating predictably at current average:
Use the predictable scale factor above to compute a hypothetical centered capability ratio:

$$Hypothetical\ C_{pk} = \frac{|\ Average - Specification\ |}{3 \times predictable\ scale\ factor} = \_\_\_\_\_\_\_\_\_\_\_\_$$

Use this value along with the *Tolerance* value found in Step 3 in the table identified in Step 5 to find a

Potential Cost of Production = ____________

Compare with the value in Step 5 to determine the benefits of operating this process predictably at the *current* average.

Part Four: Payoff for Predictable at Optimum Aim

12. Recompute the aim points:
Using the *predictable scale factor* from Step 10 find the aim points that correspond to the C_{pk} values on the next page:

$$aim\ point = specification \pm [\ C_{pk} \times scale\ factor \times 3.0\ ]$$

13. Find the nominal costs at these aim points:

14. Find the Effective Costs of Production: Use the table identified in Step 5 and *Tolerance* identified in Step 3 with the C_{pk} values shown. Multiply the Effective Costs by the nominal costs.

15. Find the potential payoff for operating predictably at optimum aim:
Compare the minimum actual cost with the values found in Step 5 to estimate the minimum payoff for operating predictably at the optimum aim point.

Worksheet for a One-Sided Specification With Target

page 6 of 6

C_{pk}	Aim Point	Nominal Cost	Effective Cost	Actual Cost
0.0	______	______	______	______
0.1	______	______	______	______
0.2	______	______	______	______
0.3	______	______	______	______
0.4	______	______	______	______
0.5	______	______	______	______
0.6	______	______	______	______
0.7	______	______	______	______
0.8	______	______	______	______
0.9	______	______	______	______
1.0	______	______	______	______
1.1	______	______	______	______
1.2	______	______	______	______
1.3	______	______	______	______
1.4	______	______	______	______
1.5	______	______	______	______
1.6	______	______	______	______
1.7	______	______	______	______
1.8	______	______	______	______
1.9	______	______	______	______
2.0	______	______	______	______

Worksheet for Target at Boundary

Part One: Characterizing Past Performance

1. **Identify a critical product characteristic:**

   Product Identification: ______________________

   Critical Characteristic Evaluated: ______________________

   Specification Limit = SL = __________

   Boundary = Target = __________

2. **Compute a *scale factor* using a global measure of dispersion:**

   Number of values used = N = ______

   scale factor: s = __________ or R/d_2 = __________

3. **Characterize the Elbow Room:**

   $$Elbow\ Room = \frac{|\ target - SL\ |}{scale\ factor} = \_\_\_\_\_\_\_\_ = \_\_\_\_\_\_\_\_$$

4. **Characterize the Distance to (Nearer) Specification:**

   average value for the N data = __________

   $$DNS = \frac{|\ average - SL\ |}{scale\ factor} = \_\_\_\_\_\_\_\_ = \_\_\_\_\_\_\_\_$$

   (*DNS* values are defined to be negative whenever the average falls on the wrong side of the specification limit.)

5. **Find the Effective Cost of Production:**

   Is nonconforming product scrapped or reworked?

   If it is reworked, then what is the cost of rework as a proportion of the nominal cost of production? 100% 80% 67% 50% 33% 20% 10%

   Choose the appropriate table from Tables 30 to 37,
   use the Elbow Room and *DNS* values found above,
   and read off your **Effective Cost of Production**: [ ]

Worksheet for Target at Boundary

page 2 of 5

Part Two: Finding the Optimum Aim Point (optional)

(If the nominal cost of production increases as you move the process average closer to the target value, then it will be of interest to determine the optimum aim point for your process:)

6. For the *DNS* values shown, find the aim points:

aim point = *specification* ± [*DNS* x *scale factor*]

DNS	Aim Point	Nominal Cost	Effective Cost	Actual Cost
0.3	______	______	______	______
0.6	______	______	______	______
0.9	______	______	______	______
1.2	______	______	______	______
1.5	______	______	______	______
1.8	______	______	______	______
2.1	______	______	______	______
2.4	______	______	______	______
2.7	______	______	______	______
3.0	______	______	______	______
3.3	______	______	______	______
3.6	______	______	______	______
3.9	______	______	______	______
4.2	______	______	______	______
4.5	______	______	______	______
4.8	______	______	______	______
5.1	______	______	______	______
5.4	______	______	______	______
5.7	______	______	______	______
6.0	______	______	______	______

Worksheet for Target at Boundary

page 3 of 5

DNS	Aim Point	Nominal Cost	Effective Cost	Actual Cost
6.3	______	______	______	______
6.6	______	______	______	______
6.9	______	______	______	______
7.2	______	______	______	______
7.5	______	______	______	______
7.8	______	______	______	______
8.1	______	______	______	______
8.4	______	______	______	______
8.7	______	______	______	______
9.0	______	______	______	______
9.3	______	______	______	______
9.6	______	______	______	______
9.9	______	______	______	______
10.2	______	______	______	______
10.5	______	______	______	______
10.8	______	______	______	______
11.1	______	______	______	______
11.4	______	______	______	______
11.7	______	______	______	______
12.0	______	______	______	______

7. **For each of the aim points above find the nominal cost of production:**
8. **Use the table from Step 5 to obtain the Effective Costs of Production:**
9. **Multiply the nominal costs by the Effective Costs to find the actual costs of production:**
   Identify the aim point that gives the minimum actual cost.

Optimum aim point = ____________

Worksheet for Target at Boundary

Part Three: Potential Payoff for Operating Predictably

10. Place your past performance data on a process behavior chart:

☐ If your process appears to be predictable, skip the remainder of this worksheet. Unless or until your process is changed in some major way, the cost of production in Step 5 is what you should expect from your process in the future. If you improve your process aim you may do as well as shown in Step 9.

☐ If your process appears to be unpredictable, then you will need to compute a *Potential Cost of Production*. As a first step toward this value use the Average Range, or the Average Moving Range, from your process behavior chart and the appropriate bias correction factor from Table 38 to compute a *predictable scale factor:*

subgroup size, n = ______ (use $n = 2$ for XmR chart)

Average Range = $\bar{R}$ = ____________

predictable scale factor = $\dfrac{\bar{R}}{d_2}$ = ____________ = ____________

11. Evaluate the benefits of operating predictably at current average:
Use the predictable scale factor above to compute hypothetical Elbow Room and *DNS* values:

$$\textit{Hypothetical Elbow Room} = \frac{|\ \textit{target} - SL\ |}{\textit{predictable scale factor}} = \_\_\_\_\_\_\_\_\_\_$$

$$\textit{Hypothetical DNS} = \frac{|\ \textit{average} - SL\ |}{\textit{predictable scale factor}} = \_\_\_\_\_\_\_\_\_\_$$

Use these values with the table from Step 5 to find a

Potential Cost of Production = ____________

Worksheet for Target at Boundary

page 5 of 5

12. Evaluate the benefits of operating predictably at optimum aim:
If Part Two of this worksheet was appropriate, then you may wish to use the predictable scale factor and the optimum aim point to compute a new hypothetical DNS value:

$$New\ Hypothetical\ DNS = \frac{|\ optimum\ aim - SL\ |}{predictable\ scale\ factor} = \_\_\_\_\_\_\_\_\_\_\_\_$$

Use this value with the Hypothetical Elbow Room from Step 11, along with the table from Step 5 to find a

Potential Cost of Production at Optimum Aim = ____________